Year 10

FOUND... N

for AQA

Keith Hirst
Mike Hiscock
David Sang
Martin Stirrup

Heinemann Educational Publishers
Halley Court, Jordan Hill, Oxford OX2 8EJ
a division of Reed Educational & Professional Publishing Ltd
Heinemann is a registered trademark of Reed Educational & Professional Publishing Ltd

OXFORD MELBOURNE AUCKLAND
JOHANNESBURG BLANTYRE GABORONE
IBADAN PORTSMOUTH NH (USA) CHICAGO

First published 2001

ISBN 0 435 57190 7

05 04 03 02
10 9 8 7 6 5 4 3 2

Edited by Teresa Brady, Tim Jackson and June Thompson

Index compiled by Paul Nash

Designed and typeset by Hardlines, Charlbury, Oxford OX7 3PS

Original illustrations © Heinemann Educational Publishers, 2001

Illustrated by Hardlines, Charlbury, Oxford OX7 3PS

Printed and bound in Spain by Edelvives.

Acknowledgements
The authors and publishers would like to thank the following for permission to use photographs:

Cover photos
Hurdler: Stone; Thermogram: Science Photo Library/Dr. Arthur Tucker; Lava flow: Science Photo Library/Bernhard Edmaier

p4: Art Directors & Trip; **p8:** Science Photo Library; **p10:** *tr* AllSport/Dave Rogers; **p12:** Mountain Camera/Pat Morrow; **p16:** *tr* Science Photo Library, *br* Science Photo Library; **p22:** John Birdsall Photography; **p24:** Science Photo Library; **p26:** *bl* Science Photo Library, *bm* Science Photo Library; **p27:** Panos Pictures/Clive Shirley 346 MOZ 735; **p30:** Science Photo Library; **p31:** Science Photo Library; **p36:** *tr* Photodisc, *mr* Science Photo Library, *br* Science Photo Library; **p40:** Science Photo Library; **p45:** Science Photo Library; **p46:** Bruce Coleman/Kim Taylor; **p48:** *tr* Science Photo Library, *mr* Science Photo Library/Saturn Stills; **p51:** Panos Pictures/Jean-Léo Dugast; **p53:** Science Photo Library; **p54:** *tr* Rex Features, *mr* Frank Lane Picture Agency/J Van Arkel (Foto Nature Stock); **p55:** Science Photo Library; **p56:** Photodisc 5062; **p58:** Photodisc 10108; **p62:** Science Photo Library; **p63:** *ml* Holt Studios, *bm* Art Directors and Trip; **p64:** Art Directors and Trip; **p73:** Science Photo Library/A Cornu T810169PDI09S **p75:** Science Photo Library/Novosti H413/029NOV066; **p76:** *tr* Peter Gould, *mr* Peter Gould, *br* Peter Gould; **p77:** Peter Gould; **p79:** Oxford Picture Library/Angus Palmer; **p80:** *tr* Art Directors and Trip/Andrew Lambert, *br* Peter Gould; **p81:** Peter Gould; **p82:** *tr* Roger Scruton, *br* Peter Gould; **p83:** Geo Science Features; **p84:** Science Photo Library; **p88:** *tr* Science Photo Library, *br* J. C. Davies; **p90:** *tr* Michael Rosenfield/Tony Stone, *mr* Julia Kamlish/Science Photo Library, *br* Beyen of Cowes; **p91:** *tr* Eye Ubiquitous/Paul Seheult, *mr* Tony Gudgeon; **p92:** Peter Gould; **p93:** Art Directors and Trip; **p95:** Nigel Cattlin/Holt Studios; **p100:** Environmental Images/Pete Addis; **p102:** Frank Spooner/Chip Hires; **p110:** *mr* Amanda Gazidis/Environmental Picture Agency, *br* Tony Gudgeon; **p112:** *mr* S Whitehorn/Environmental Picture Agency, *br* David Townend/Environmental Picture Agency; **p114:** Art Directors and Trip; **p116:** *tr* Art Directors and Trip, *br* Jim Greenfield/Planet Earth Pictures; **p118:** Landform Slides; **p119:** (both pics) Geo Science Features Picture Library; **p120:** *tr* Science Photo Library, *mr* Frank Lane Picture Library; **p128:** James Davies Travel Photography; **p130:** Panos Pictures; **p136:** Telegraph Photo Library/Bryan and Cherry Alexander; **p137:** Environmental Images/Martin Bond; **p139:** Environmental Images/Chinch Gryniewicz; **p144:** Peugeot; **p146:** Rex Features/Simon Roberts; **p148:** *tr* The Electricity Board, *br* Garden & Wildlife Matters Photo Library/J Feltwell; **p150:** Peter Gould; **p152:** *tr* Corbis, *br* Eye Ubiquitous/Paul Thompson; **p153:** Holt Studios/Inga Spence; **p154:** Oxford Scientific Films/Breck P. Kent; **p155:** Science Photo Library; **p156:** *tr* BNFL, *br* Science Photo Library; **p158:** *tr* Art Directors and Trip, *mr* Art Directors and Trip; **p159:** Environmental Images/Martin Bond; **p160:** Science Photo Library; **p162:** *tr* Oxford Scientific Films/NASA, *br* Ecoscene/John Farhar; **p164:** Environmental Images/David Lawson; **p169:** Science Photo Library H407/084JBU01I; **p170:** *tr* Science Photo Library, *br* Peter Gould; **p171:** Frank Spooner/Rich Gigle Liai; **p172:** Janine Wiedel; **p173:** Science Photo Library; **p174:** Janine Wiedel; **p176:** Science Photo Library; **p177** Science Photo Library; **p179:** Janine Wiedel; **p180:** *tr* What PC, *mr* Janine Wiedel; **p186:** Repfoto/Robert Ellis; **p188:** *tr* Peter Gould, *mr* Art Directors and Trip, *mb* Art Directors and Trip, *bl* Garden Matters; **p189:** Art Directors and Trip; **p192:** *mr* Art Directors and Trip, *tr* Telegraph Colour Library/FPG Navaswan Industrial; **p193:** *m* Art Directors and Trip; **p195:** Science Photo Library; **p196:** Peter Gould; **p200:** *tm* Peter Gould, *tr* Art Directors and Trip; **p202:** *tr* Science Photo Library, *mr* Peter Gould.

Picture research by Jacqui Rivers

The authors and publishers would like to thank the following for permission to reproduce copyright material:

p18: The Sunday Times, Times Newspapers Ltd.

The publishers have made every effort to trace the copyright holders, but if they have inadvertently overlooked any, they will be pleased to make the necessary arrangements at the first opportunity.

About this book

This is the first book in the AQA Modular Science series and is designed to guide you through the first year of GCSE study. It has been written to help you find everything you need to pass your module tests and prepare for your exams.

We have named each module in the book with the title of the module test that you will be set. This is so that you can be sure of what you need to study.

The introductory page to each module will help you get a feel of the module. It explains the topics covered in the module and gets you thinking about the science that you already know.

To help you study we have included some useful features in the book. Here are a few:

Double-page spreads
Everything you need to know for each module is covered in double-page spreads. These pages will cover all the topics you need to understand for your module tests. To help you find all of the important points we have included questions to test yourself as you learn.

End of module questions
At the end of each module you will find three pages of test questions. These questions are similar to the type of questions you will find in your end of module tests. Answering lots of questions will help you check what you have learnt and prepare for the tests.

Did you know?
These boxes contain extra information about the topic that you are studying. The information in these boxes will not be tested in module tests or exams.

Summary
The information in these boxes summarises the most important points on the page. In your module tests you will be tested on your knowledge of the points in the summary boxes. These boxes will also help you make notes and answer questions.

Glossary pages
When a new scientific word appears for the first time in the text, it will appear in **bold** type. All words in bold are listed with their meanings in the glossary at the back of the book. Look there to remind yourself what they mean.

Data sheet pages
When you sit your module tests and exams you will be given data sheets. At the back of the book you will find the information that will be on these data sheets.

Contents

Module 1 – Humans as organisms

Teenage obesity soars in the UK

THE YOUNG RISK THEIR HEALTH

Parents urged to trust MMR vaccine

Energy drink claims – a con!

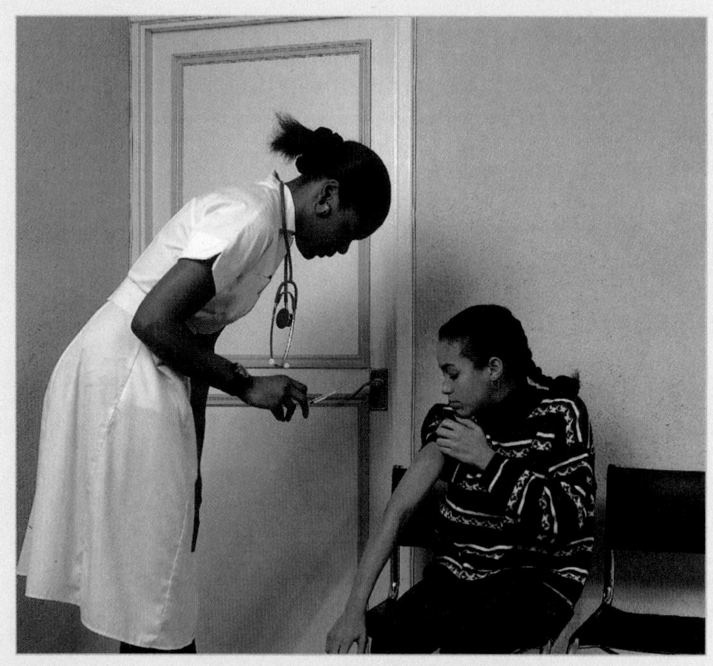

These are just some of the stories hitting the headlines. Stories about health issues are constantly in the newspapers, magazines, and on the TV. Advertising campaigns about health products can be very persuasive. We are told what we should and shouldn't eat and drink, the kind of exercise we should take, how to stay healthy when we go on holidays, and whether it is safe to be vaccinated. So how do you make important health decisions?

Humans as organisms will help you to understand what is involved in many of these health issues. By covering topics such as 'what happens to the food you eat?', 'how your heart works', 'how your body responds to exercise', and 'how disease can be prevented', this module will explain the science background to many of the health issues which you face now and in your future.

Try these questions first.

1. Where do you get the energy you need to perform your life processes?
2. What are the main organs that make up your digestive system?
3. What things do you need to do in order to keep your body healthy?
4. How does your body defend itself to fight infection?

1:1 Body building

Living building blocks

Your body is made from millions of tiny cells. Just as a house is built from bricks your body is built using cells.

Most human cells have:

- a nucleus;
- cytoplasm;
- a cell membrane.

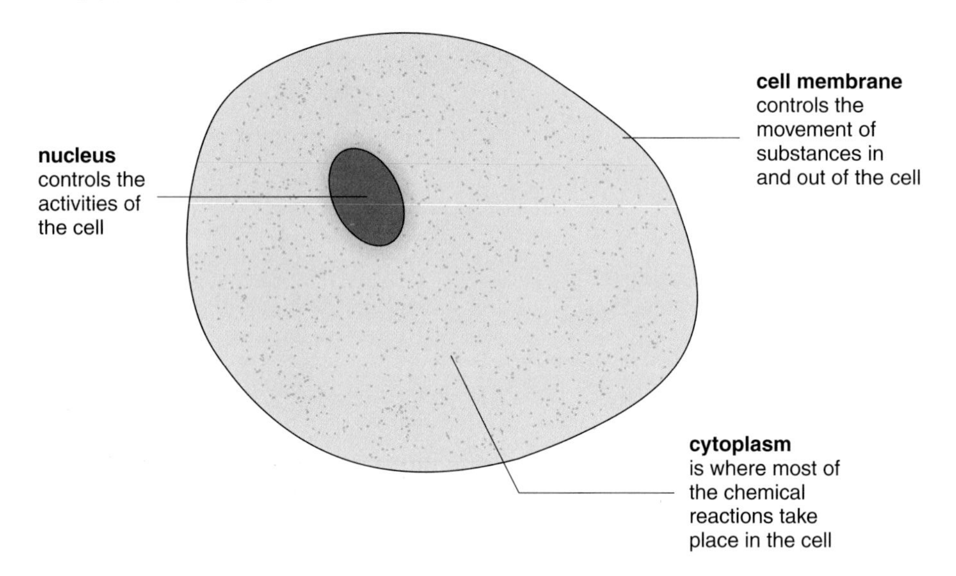

Parts of a cell.

Specialised cells

The cells of your body have different shapes and sizes. This is because the structure of cells makes them suited to do their job. Here are a few examples:

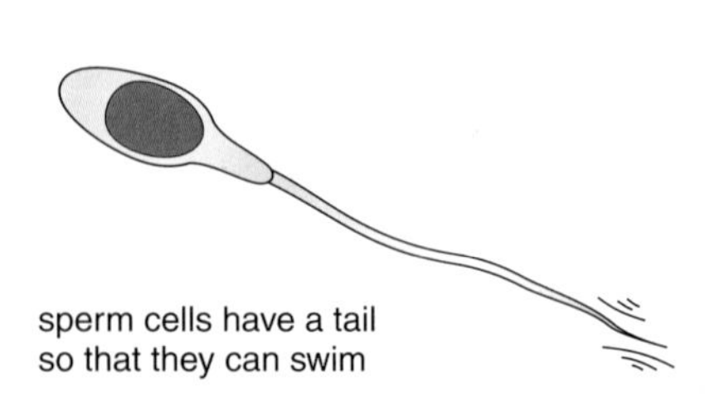

sperm cells have a tail so that they can swim

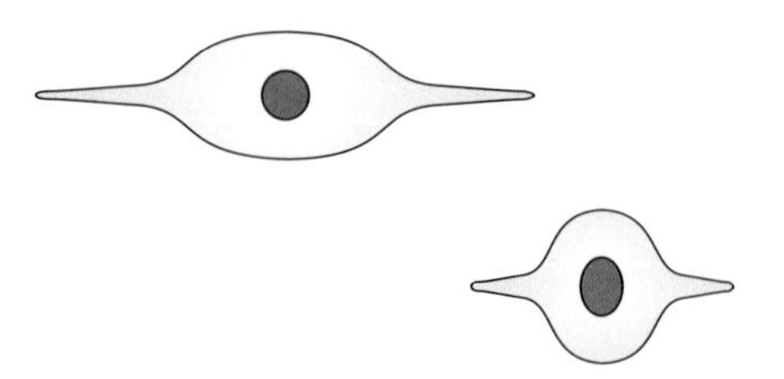

muscle cells are able to contract so that they can become shorter

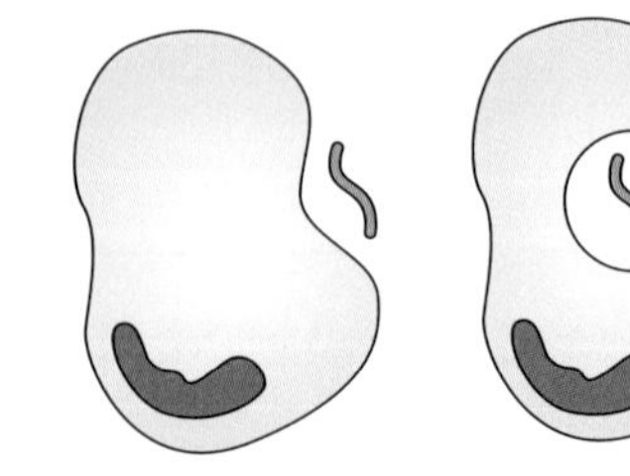

white blood cells can change shape to destroy bacteria which can infect your body

a Copy and complete this table:

Type of cell	Special feature	What feature does
sperm cell		
white blood cell		
muscle cell		

Working together

A group of cells with a similar structure working together to carry out the same job is called a **tissue**. For example, muscle tissue is made from many muscle cells working together.

Different types of tissues group together to build the **organs** of your body. For example, your stomach is an organ made from different tissues.

Different organs group together to form the **organ systems** of your body such as your digestive system and your circulatory system.

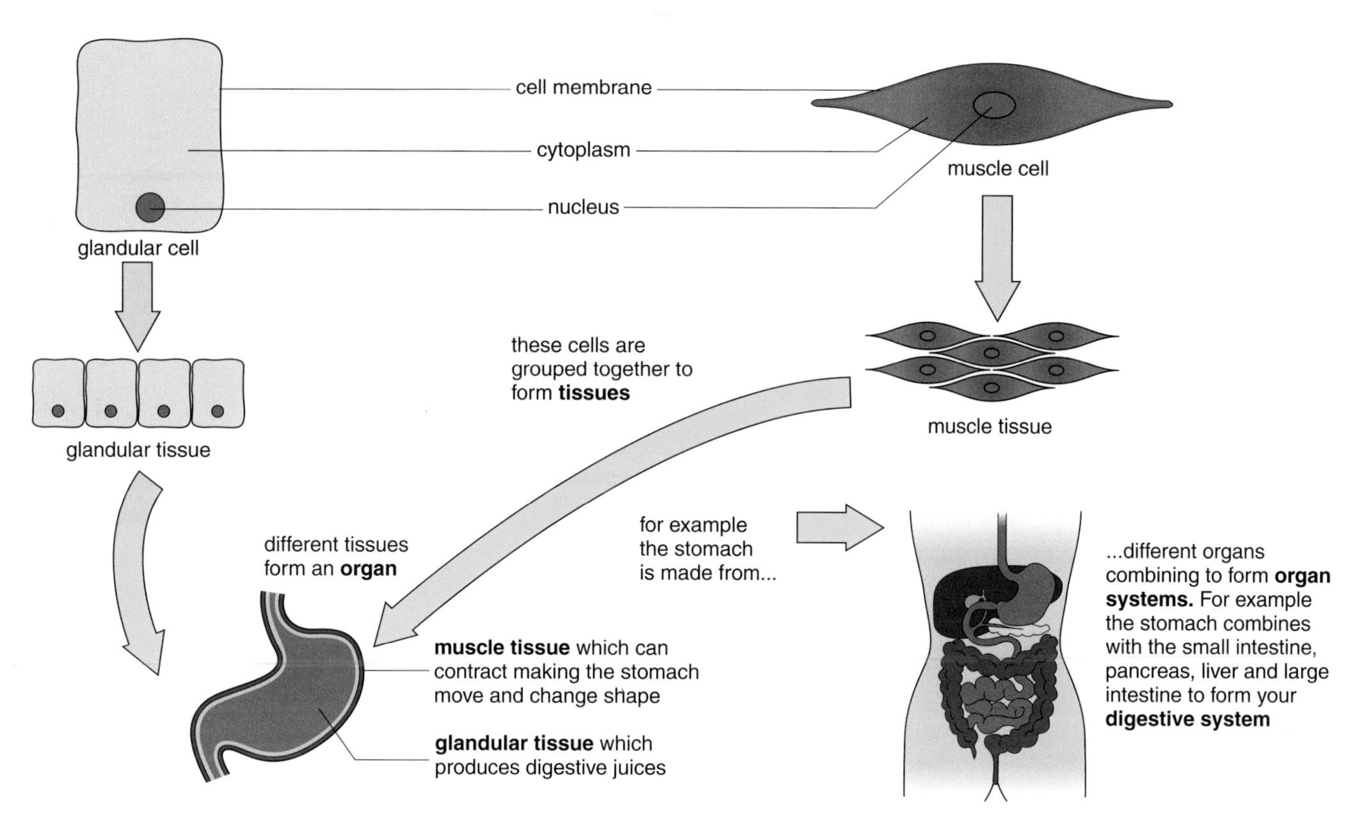

Questions

1 **Copy and complete this table:**

Cell part	What it does
nucleus	
cytoplasm	
cell membrane	

2 **Draw a muscle cell and label the nucleus, cytoplasm and cell membrane.**

3 **Copy and complete the sentences below using the words provided.**

membrane   gland   tissue   stomach   tissues

A group of cells with the same shape doing the same job is called a __________. An example of an organ system is __________. Each organ system is made from different __________ working together.

Summary

- Your body is made from cells.
- Most cells have a nucleus, cytoplasm and a cell membrane.
- Cells are different so that they can do different jobs.

1:2 Cutting food down to size

Getting food into the body

The patient in the photograph opposite is getting the food she needs through a drip connected directly to her bloodstream.

The drip contains soluble sugars which can dissolve in her blood. Once the sugars are in the bloodstream they are transported around her body, providing her cells with the energy and nutrients she needs to recover.

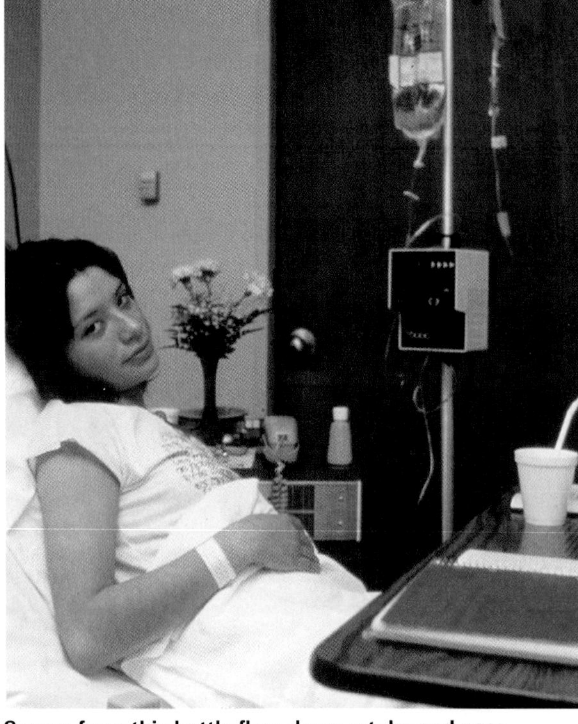

Sugars from this bottle flow down a tube and pass directly into the girl's bloodstream.

Making foods usable

If you are not ill and can eat normally, you usually get your energy and nutrients from the food in your diet.

Most of the foods you eat are made from large, insoluble molecules. These have to be broken down into small, soluble molecules which can then dissolve in your blood and be carried around your body.

For example, many foods contain starch which is a large, insoluble molecule. Starch is broken down (digested) into sugars which are small, soluble molecules.

Your **digestive system** converts large, insoluble molecules into small, soluble molecules.

a **Explain why hospital drips contain only soluble substances.**

Small molecules

Your diet contains the three types of food – **carbohydrates, proteins** and **fats**. Each type of food is insoluble and has to be digested into small, soluble molecules.

The table opposite shows what is produced when each type of food is digested.

Type of food	Product of digestion
carbohydrate	sugars
fats	fatty acids and glycerol
protein	amino acids

b **A cheese sandwich contains protein (the cheese), starch (the bread) and fat (the butter). What is produced from each of these foods when they are digested?**

Breaking down your food

Your digestive system produces **enzymes**. These **catalyse** (or speed up) the breakdown of larger molecules into small ones. The diagram opposite shows how enzymes work to digest food.

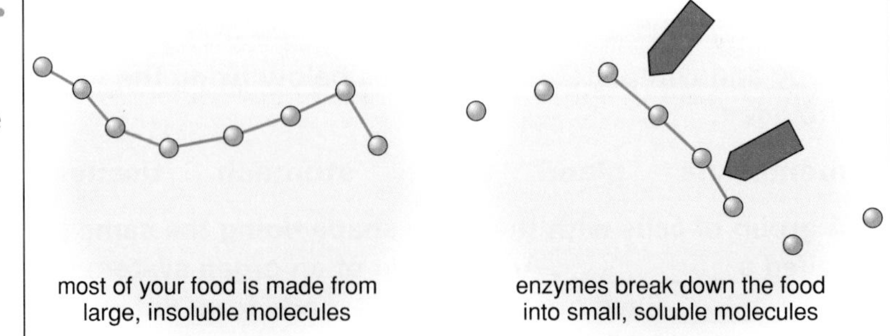

The right tools for the job

Each type of food needs a particular enzyme to break it down. For example, **protease** enzymes digest protein into amino acids.

The enzymes which catalyse the breakdown of carbohydrates into sugars are called **carbohydrases**.

Lipase enzymes catalyse the breakdown of fats into fatty acids and glycerol.

The diagram below shows how enzymes work to digest your food into small, soluble molecules which can be absorbed into your body and dissolve in your bloodstream.

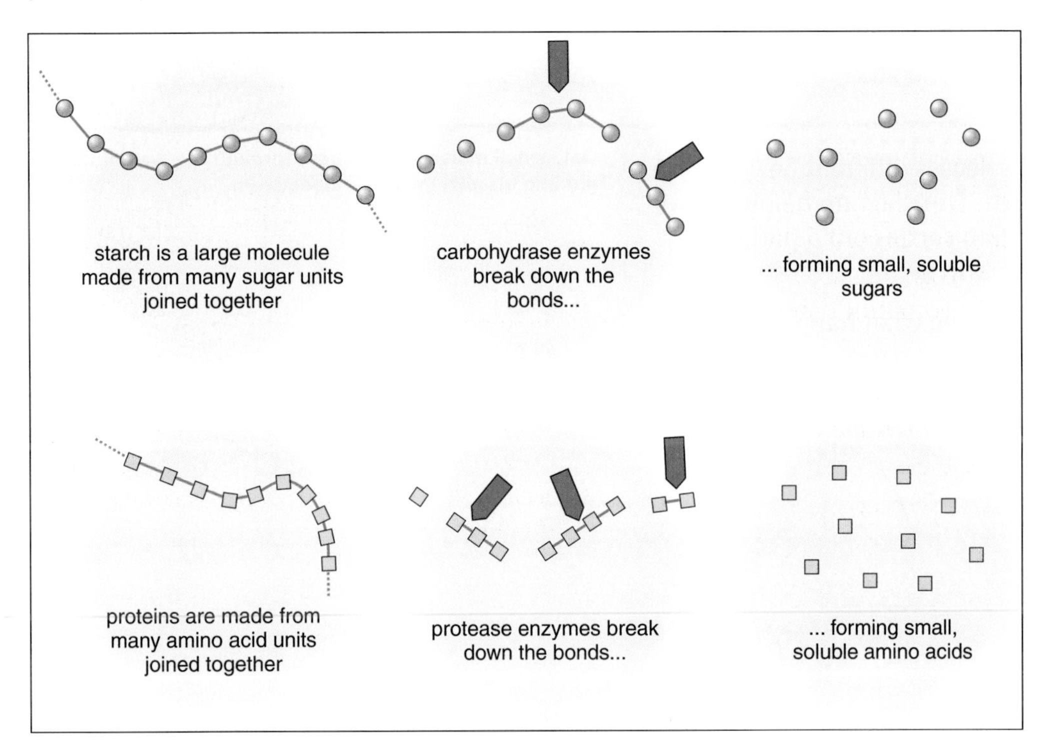

Questions

1 Copy and complete this table:

Type of food	Type of enzyme	Digestion products
starch		sugars
	lipase	
		amino acids

2 Explain why proteins in the food you eat need to be broken down by enzymes.

3 Explain why health drinks, such as 'Lucozade', contain sugar.

Summary

- Starch, protein and fats are insoluble.
- During digestion large, insoluble molecules are broken down into small, soluble molecules.
- Only small, soluble molecules can be absorbed into the bloodstream.
- Digestion is speeded up by enzymes.

1:3 Speeding things up

Making big molecules smaller

Much of the food you eat is made from large, insoluble molecules. Only small molecules can pass from your gut into your bloodstream.

Large food molecules have to be broken down (digested) as they pass along the digestive system.

a Why does your food need to be broken down?

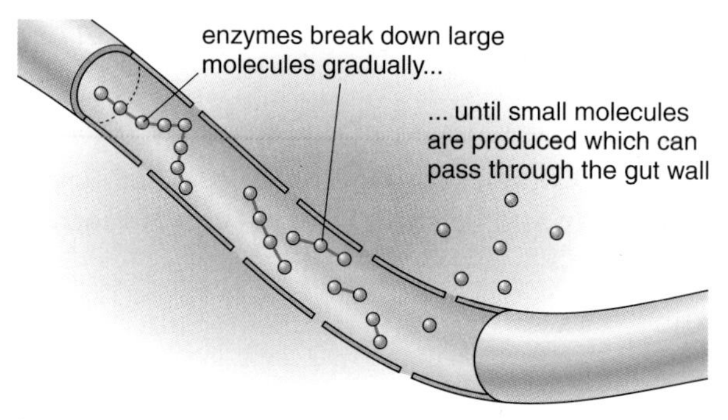

Only small molecules can pass through the wall of the gut and dissolve in your bloodstream.

Stages of digestion

Each part of the digestive system releases different enzymes on to the food you eat to break it down. The diagram below shows how each part of the digestive system carries out a particular role in the gradual breakdown of food.

Study the diagram carefully by following what happens to food as it passes from the mouth to other parts of the digestive system.

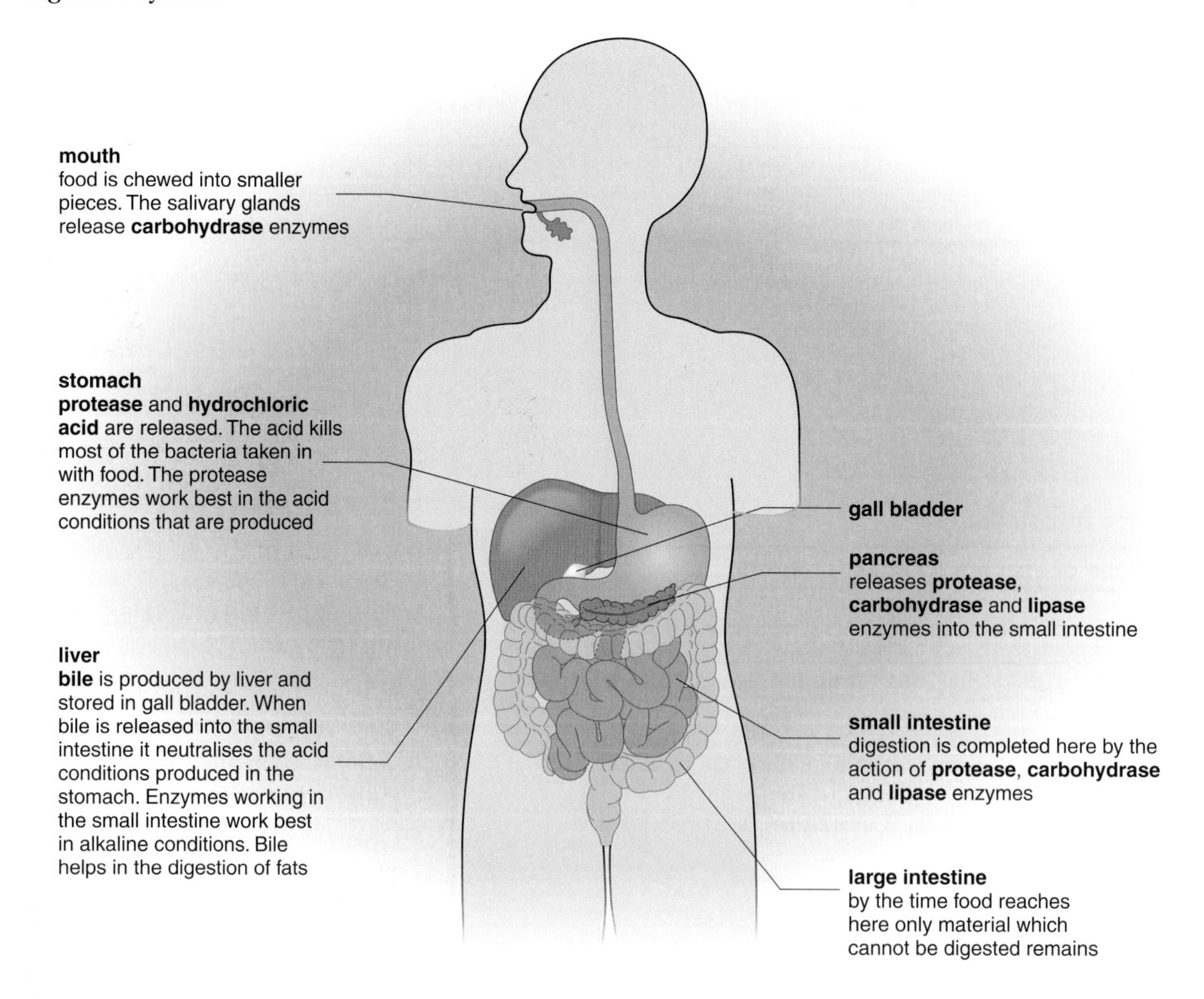

Food is gradually broken down as it passes along the gut.

Big drops to little droplets

Bile is released from your liver on to food in your small intestine.

When bile mixes with food it neutralises the acid produced by the stomach and makes the conditions in the small intestine alkaline. All the enzymes released in the small intestine work best in alkaline conditions.

Bile also helps to digest fats. It does this by breaking down large fat drops into thousands of tiny fat droplets. This process is called **emulsification**. The surface area of fat droplets is much greater than the surface area of the large drop of fat. This increases the surface area of fats for lipase enzymes to act upon and so speeds up fat digestion.

The diagram opposite shows how emulsification works.

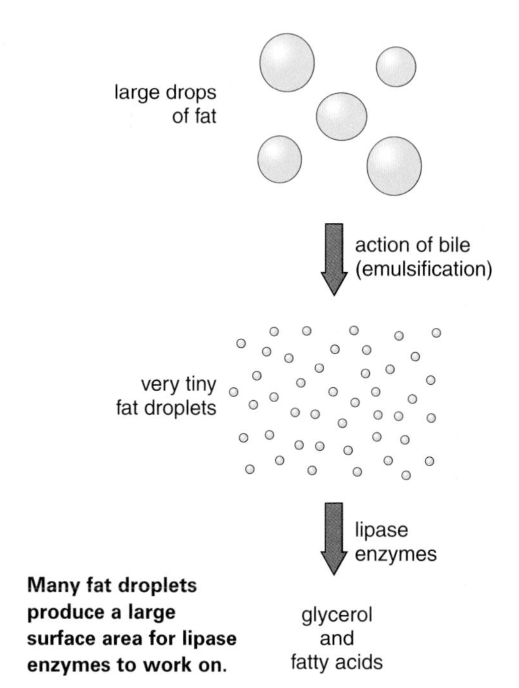

Many fat droplets produce a large surface area for lipase enzymes to work on.

b What sort of chemicals help to break down your food during digestion?

Investigating enzyme action

A group of students carried out an investigation using a protease enzyme from the gut. The students also used a protein called egg white.

The protein starts off looking cloudy. If all the protein is digested the solution will go clear. The diagram opposite shows the results of their investigation. Study the diagram carefully and answer the questions below.

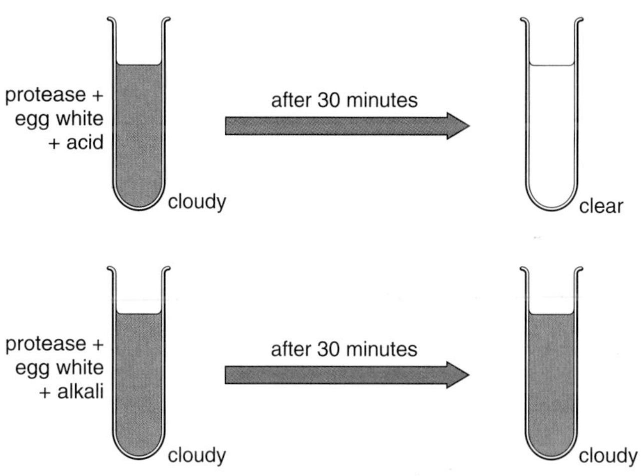

c In which conditions does the protease work best?

d In which region of the gut is this protease likely to be found?

Questions

1 Copy and complete the following sentences using words from this list:

liver stomach small intestine fats starch protease lipase

Bile is produced in the __________ and is added to food in the __________. When bile is added to food it emulsifies __________ so that __________ enzymes can work better.

2 Where in the body is bile produced?

3 What is the main job of bile?

Summary

- Each part of the digestive system carries out a particular job in digesting food.
- As well as producing enzymes, digestive organs also produce acid or alkaline substances so that enzymes work effectively.
- The liver produces bile which neutralises the acid added to food in the stomach.
- Bile also emulsifies fats.

1:4 Getting food into the body

What happens to digested food?

Once food molecules have been broken down into small, soluble molecules the process of digestion is complete. Molecules of digested food are small enough to be absorbed through the gut wall. The molecules are also soluble so that they will dissolve in the bloodstream which then transports the food to all parts of the body.

The surface of the small intestine is covered by villi which increase surface area to speed up absorption.

Getting food into the body

Your small intestine is very efficient in absorbing food because its structure is specialised to carry out this job. The small intestine is able to absorb large amounts of food because:

- its inner surface contains thousands of tiny folds called **villi** which produce a very large surface area for food molecules to pass through;
- each villus contains many tiny blood vessels (capillaries) to transport absorbed food;
- each villus is very thin so that food molecules can easily reach the bloodstream.

The diagram below shows how these features allow absorption to take place very efficiently.

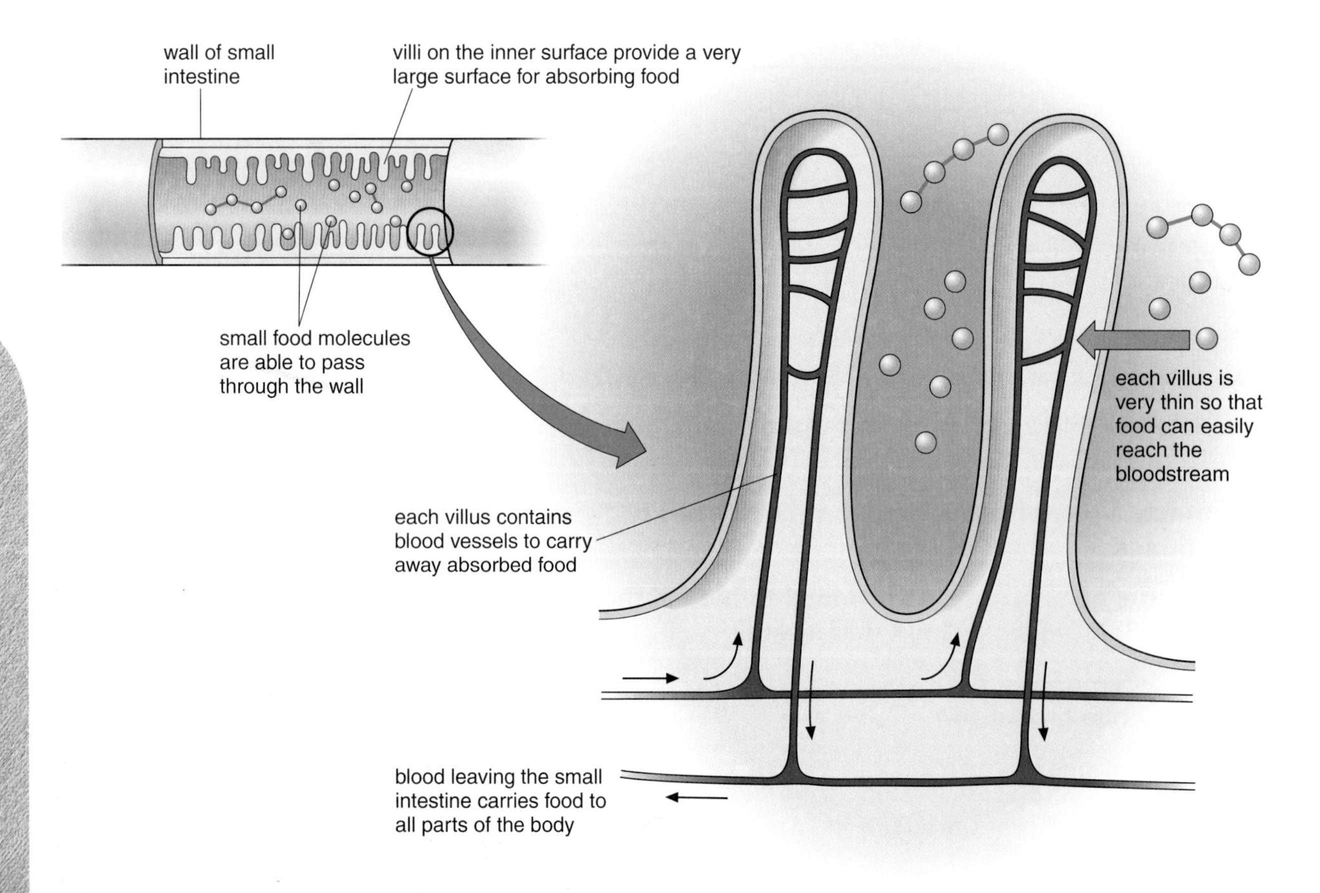

Getting food to all parts of the body

Once food molecules reach the bloodstream they dissolve in **plasma** – the liquid part of blood. Plasma containing dissolved foods is then transported to all parts of the body.

This is why it is important to have such a large number of blood capillaries in the wall of the small intestine.

a **List three food substances which are likely to be found in blood plasma as it leaves the small intestine.**

Getting rid of waste

When food reaches the **large intestine** all the soluble foods will have been absorbed, leaving behind substances which cannot be digested, such as **fibre** (cellulose).

Even though you cannot digest the fibre in your diet it is still beneficial. Fibre helps to prevent constipation and diseases of the intestine, including cancer.

In the large intestine much of the water is absorbed from the undigested material leaving behind semi-solid waste called **faeces**. This waste is then removed from the body through the anus.

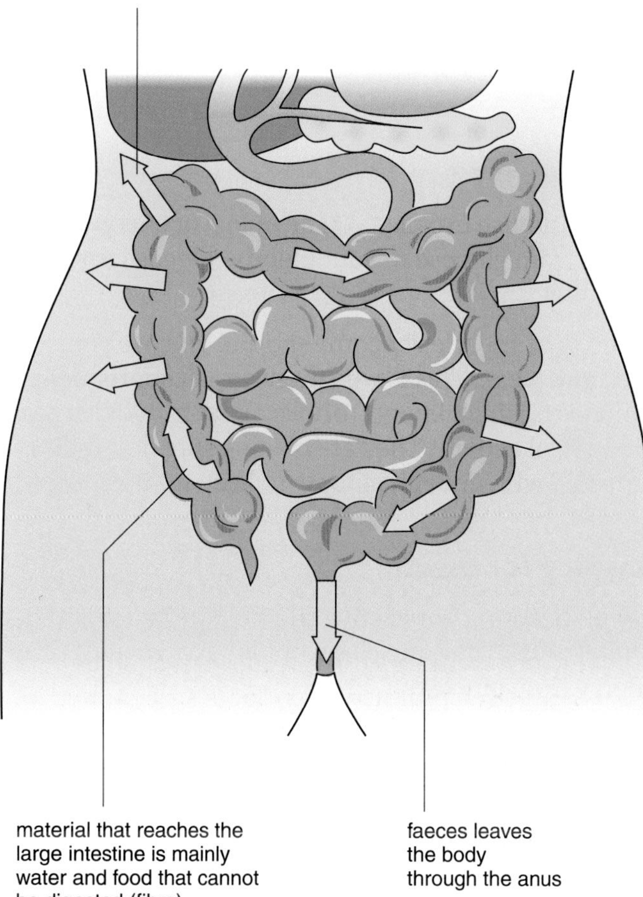

The large intestine absorbs water and forms waste (faeces).

Question

1 **Copy and complete the following text, using words from the list below.**

villi   blood vessels   specialised   folds

The structure of the small intestine is __________ to carry out is job. The inner surface contains thousands of tiny __________ called __________. Each villus contains tiny __________ __________ to transport absorbed food.

2 **Which part of the bloodstream will contain dissolved food?**

3 **What is the name given to food substances that cannot be digested? Why is it important to have this in your diet?**

Summary

- Digested foods can be absorbed through the wall of the small intestine.
- The small intestine has special features which make it a very efficient organ for absorbing food.
- The large intestine absorbs water from the indigestible food which forms faeces.

1:5 Releasing energy

Keeping going

During intense exercise you need lots of energy. This energy comes from glucose in your blood – glucose is your body's main fuel.

When your muscles become short of glucose you start to feel tired and weak. This is why even very fit sportsmen and women use 'high energy' drinks or eat fruit during long periods of physical activity. They are making sure that their muscles are provided with enough glucose to keep up their energy levels.

Sportsmen and women use 'high energy drinks' to keep up their energy levels.

Gasping for breath

As you become more active you breathe faster and deeper. This is to provide your muscles with the oxygen needed to release energy from glucose.

Oxygen and glucose react together in body cells to release energy. This process is **aerobic respiration**. Activities such as running need lots of energy from respiration. Aerobic respiration is summarised by the equation:

glucose + oxygen → carbon dioxide + water + energy

Using energy

The energy that is released during respiration is used by your body to:

- build larger molecules from smaller ones, so that you can grow and repair damaged cells;
- enable muscles to contract so that you can move;
- keep your body temperature steady even when you are in colder surroundings.

Respiration without oxygen

Sometimes your body cannot get enough oxygen. For example, during very strenuous exercise such as sprinting, your muscles are working so hard that they become short of oxygen. To provide the energy that is needed, glucose is broken down without using oxygen. This process is called **anaerobic respiration**. This process produces **lactic acid** as a waste product instead of carbon dioxide and water. Anaerobic respiration is summarised by the equation:

glucose → lactic acid + energy

Getting into debt

The lactic acid produced by anaerobic respiration has to be removed. This is done by using oxygen to break down lactic acid. After exercise you often keep breathing heavily to take in extra oxygen. This is because during anaerobic respiration you build up an **oxygen debt** and when the exercise is over you have to repay the debt by taking in extra oxygen.

Questions

1

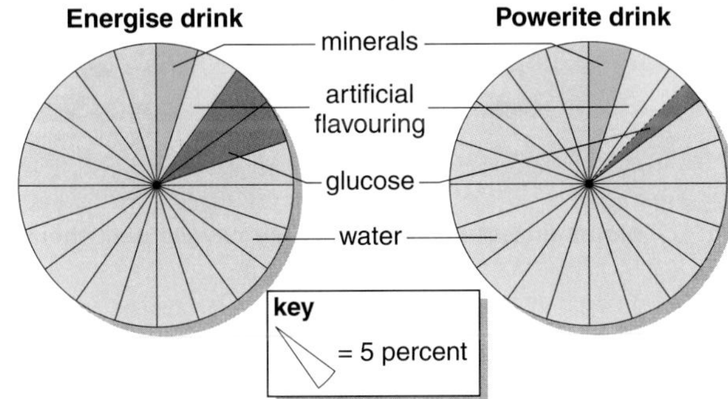

a Using the data in the pie chart copy and complete the following table:

	Energise	Powerite
% water		
% glucose		
% minerals		
% flavouring		

b Which of the substances in the drinks provides energy to the body?

c Which of the two drinks is the best to take when resting during a sports match? Give a reason for your answer.

2 Copy and complete the table:

	Substances needed	Substances produced
aerobic respiration	glucose and ________	water and ________ ________
anaerobic respiration	________	________ ________

Summary

- In aerobic respiration glucose and oxygen react to release energy.
- Carbon dioxide and water are produced during aerobic respiration.
- Energy is released from glucose in anaerobic respiration without using oxygen.
- Lactic acid is produced during anaerobic respiration.
- The oxygen needed to break down lactic acid is called an oxygen debt.

1:6 Take a deep breath

Climb to the top

We all need oxygen to keep us alive. In a healthy body the breathing system is working all the time to provide a continuous supply of oxygen.

The air around very high mountains contains less oxygen. This is why mountaineers have to take a supply of oxygen with them.

Fortunately, the air you breathe contains sufficient oxygen to keep you fit and healthy.

Mountaineers need to take their oxygen with them.

Air passages

The job of your breathing system is to get air into and out of your body. This brings fresh supplies of oxygen into the lungs and then removes carbon dioxide. The diagram below shows your breathing system. Study the diagram carefully. Notice the three main parts:

- a series of **air passages** which connect your lungs with the air outside;
- gas exchange tissues made from millions of tiny air sacs called **alveoli**;
- the ribcage made from bone and muscle which protects your lungs and is used to pump air in and out of your lungs.

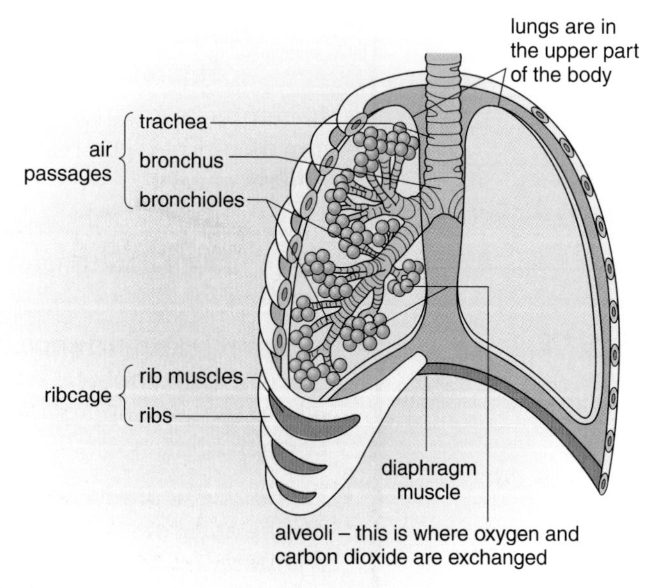

The breathing system.

Getting air in and out

The diagrams below show how your breathing system works. The movement of air in and out of the lungs is called **ventilation**. Study the diagrams carefully to find out how air gets into your lungs and then out again.

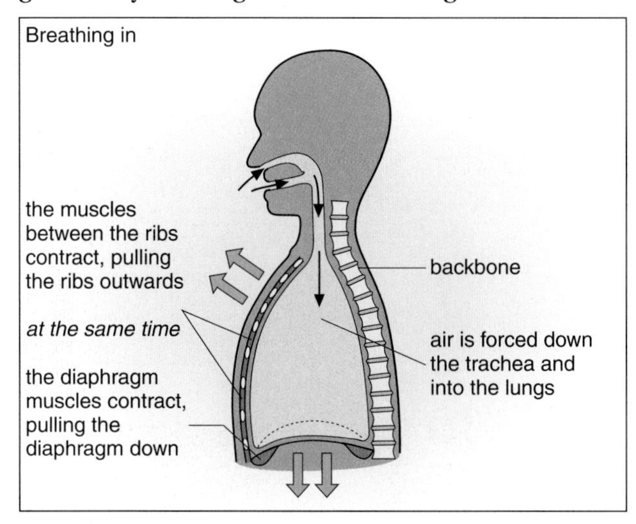

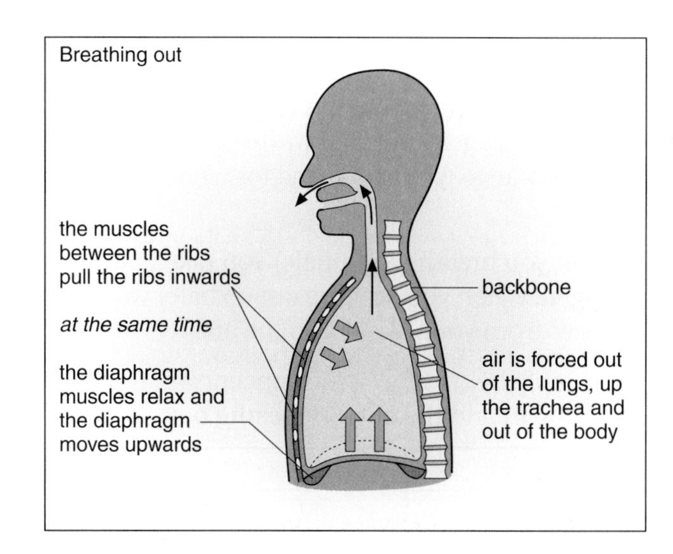

Investigating breathing

As you become more active your cells need more energy. To provide more energy you need to break down more glucose during respiration. This means that you need to get more oxygen into your body. The table below shows the results of an investigation carried out to find out how breathing changes during exercise.

Activity	Volume of each breath (cm^3)	Number of breaths taken per minute
rest	500	18
20 step-ups per min	750	25
50 step-ups per min	1200	34

a **How does breathing change as activity becomes more demanding?**

Questions

1 **Copy out and complete the following sentences, using the words provided.**

body ribs contract ventilation diaphragm

. **When a person breathes in the __________ moves out and the __________ becomes flatter.**

2 **Use the results from the breathing investigation to answer the following:**

a **How many breaths did the person take per minute at rest?**

b **How many more breaths were taken when the person was doing 20 step-ups?**

Summary

- In the lungs oxygen passes into the bloodstream and carbon dioxide passes into the air.
- Movement of the ribs and diaphragm makes air move in and out of the lungs.
- Movement of air in and out of the lungs is called ventilation.

1:7 Exchanging gases

A fair swop

As you breathe you are not only taking in the oxygen you need, you are also getting rid of waste carbon dioxide. In your alveoli you exchange what you need for what you don't need – your waste.

When you breathe in (inhale) you take in air containing a lot of oxygen. When you breathe out (exhale) waste carbon dioxide moves from your blood into the air.

The table below shows the concentration of gases in the air you breathe in and the air you breathe out.

Gas	Inhaled air (%)	Exhaled air (%)
oxygen	20	16
carbon dioxide	0.04	4
nitrogen	78	78

a Describe the differences in the amount of each gas shown in the table.

Spreading out

The exchange of oxygen and carbon dioxide in your lungs takes place because of **diffusion**.

The molecules of a gas such as oxygen are moving about all the time. Because they are moving about the molecules spread themselves out evenly. This process is called diffusion.

The effect of diffusion is that molecules move from where there is a high concentration to where there is a low concentration.

The rate of diffusion is fast when there is a large difference in concentration.

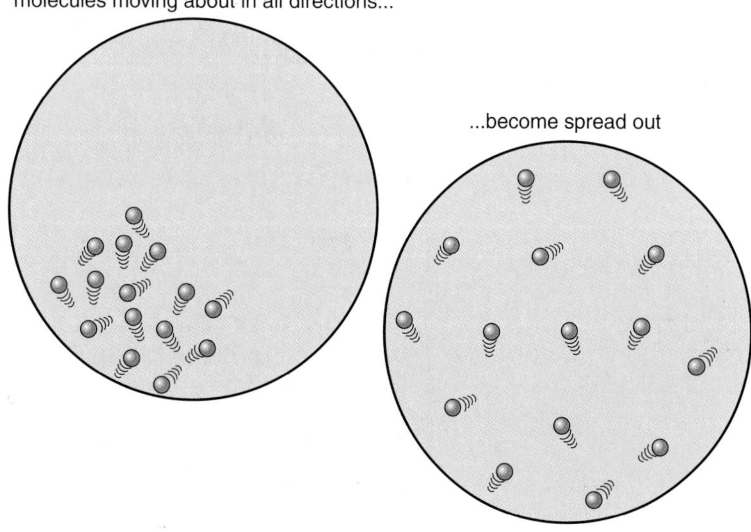

Exchanging gases

Once the air you breathe in reaches the alveoli, oxygen and carbon dioxide are exchanged by diffusion. The diagram on the right shows how the oxygen you need diffuses into the bloodstream and how waste carbon dioxide diffuses out of the blood.

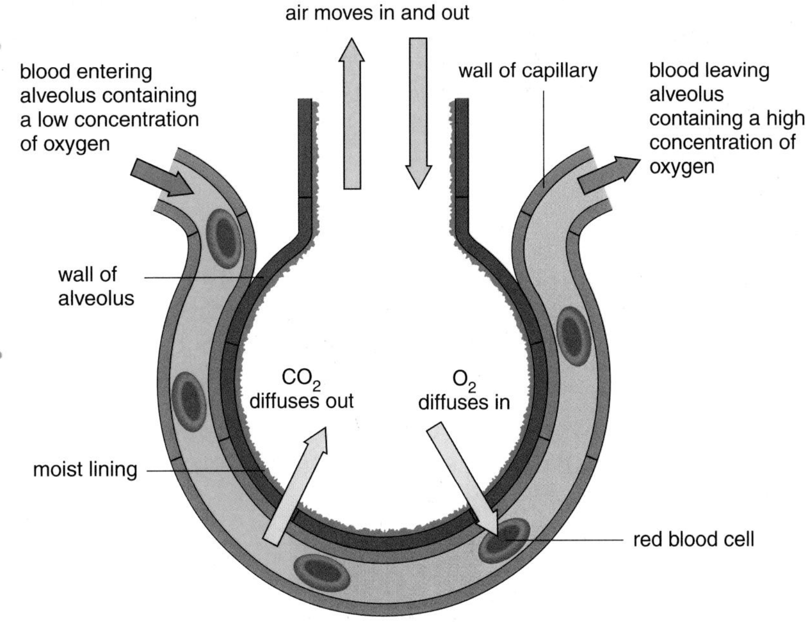

Close proximity of alveoli and blood vessels allows gas exchange by diffusion.

Damaging the lungs

When someone smokes a cigarette harmful substances are inhaled into the air passages and alveoli. This is why smoking damages the lungs. For example, damage to the alveoli can result in **emphysema** – a condition in which the walls of the alveoli break down so reducing the surface area for gas exchange. The diagram below shows this effect of emphysema.

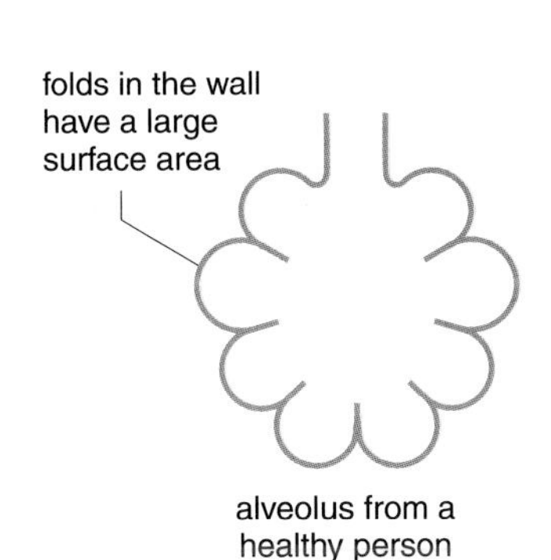

damage to the wall reduces surface area

alveolus from a person with lung damage (emphysema)

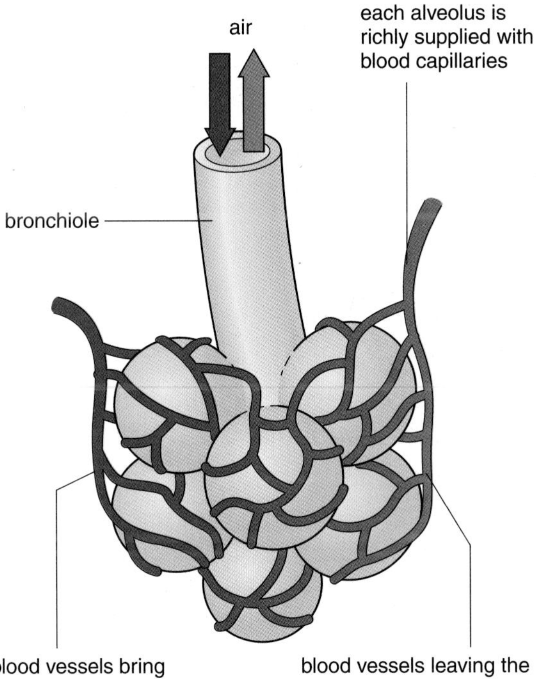

Alveoli are richly supported with blood vessels.

Questions

1 Use words from the list to complete these sentences.

carbon dioxide   digestion   oxygen   respiration   breathing   diffuses

The air entering the alveoli contains a high concentration of __________. This gas __________ into the bloodstream and is transported to all body cells and used in __________.

2 Alveoli increase the surface area for gas exchange in the lungs. Explain what happens when the alveoli become damaged by cigarette smoke.

Summary

- Substances diffuse from a high to a low concentration.
- Oxygen diffuses from the air in the alveoli into the bloodstream.
- The alveoli increase the surface area for gas exchange.

1:8 Supplying body needs

Life blood

Your blood provides all the cells of your body with the materials they need as well as removing waste substances. Cells would soon stop working without a good blood supply.

If people have lost a lot of blood in an accident they can be given blood that has been donated from someone else. The blood being given to the patient in the photograph will help to save his life.

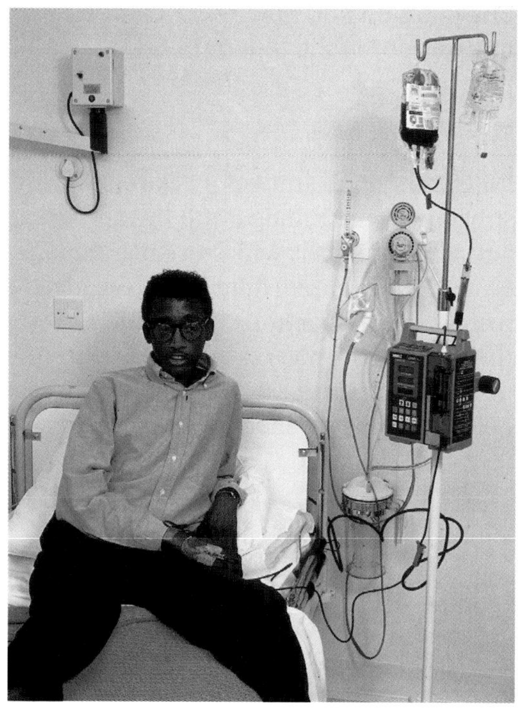

Blood donors save lives.

What is in your blood?

Your blood contains a large number of cells floating in a liquid called plasma. The diagram and photograph below show what the cells in your blood look like and what they do.

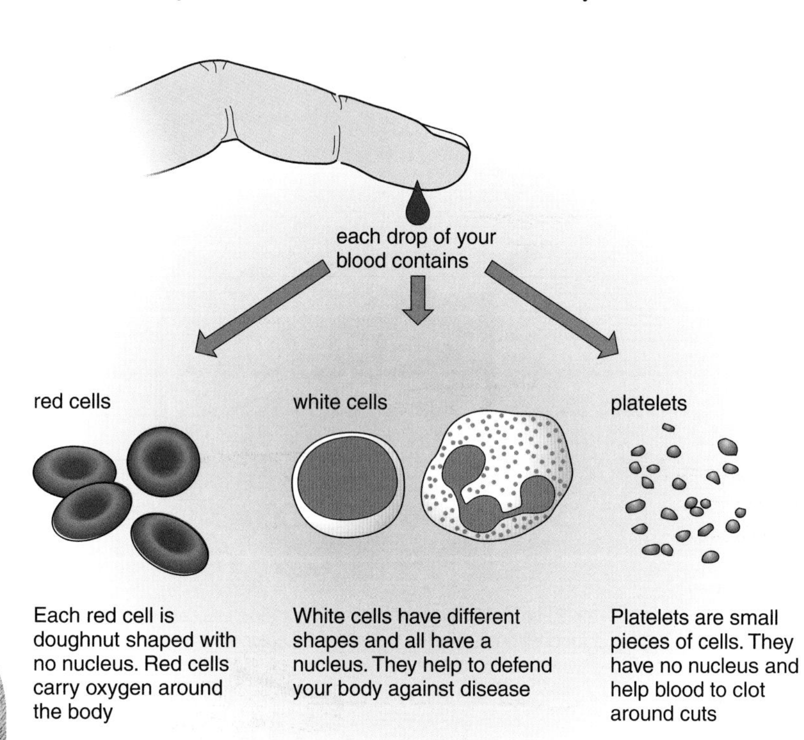

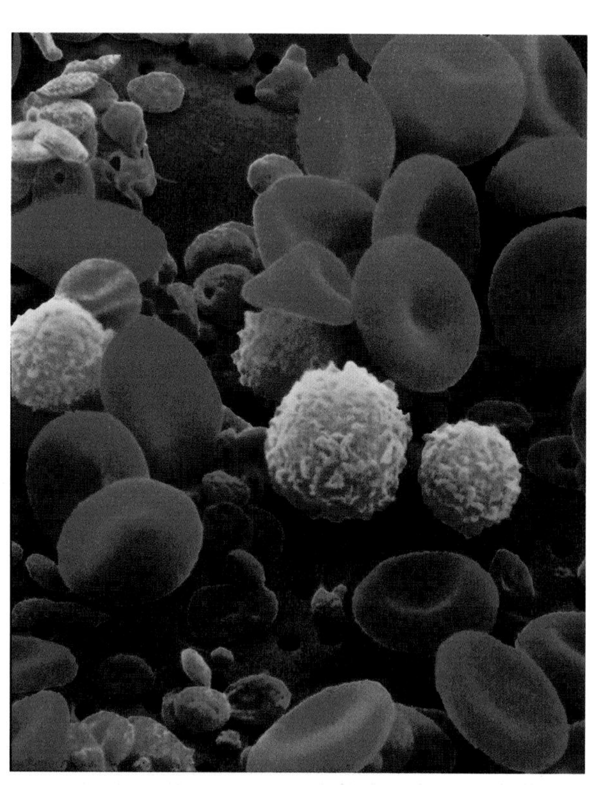

A photograph of blood cells taken using a microscope.

Carrying oxygen

Blood contains an enormous number of red blood cells. Red blood cells transport oxygen from your lungs to all the cells of your body. Cells then use the oxygen to release energy during respiration.

You can see in the diagram at the top of page 17 that red blood cells combine with oxygen as blood flows through the lungs. Oxygen is released from red blood cells as blood flows through body tissues.

Transporting useful materials

Plasma is the liquid part of blood. It contains mainly water with many substances dissolved in it. These soluble substances are transported in plasma from one part of your body to another.

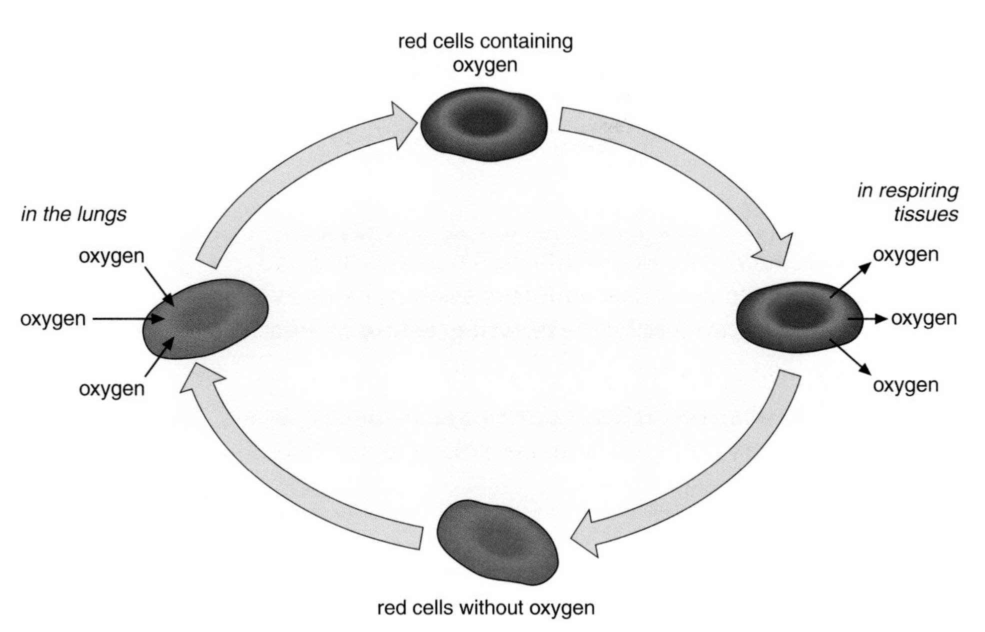

The main substances transported around your body in plasma are:

- *carbon dioxide* which is produced by all body cells as they respire. The carbon dioxide produced by body cells is carried in plasma to your lungs and then breathed out;
- the *soluble products* of *digestion* which are absorbed from the small intestine and transported in plasma to other body organs;
- *urea* which is a waste substance produced in your liver. It is transported in plasma from the liver to your kidneys and removed from your body in urine.

Questions

1 List three substances transported in blood plasma.

2 Copy and complete the following table:

Type of blood cell	Appearance/function
	transports oxygen around the body
	helps the blood to clot
	contains a nucleus
	are biconcave in shape
	defends the body against disease

3 Copy and complete the following table by writing true and false in the spaces.

Plasma transports ...

oxygen from the lungs to muscle	
starch from the small intestine to muscle	
carbon dioxide from muscles to the lungs	
carbon dioxide from the lungs to muscle	
urea produced by the kidneys to the liver	

Summary

- Blood transports useful and waste substances around the body.
- Blood contains red blood cells, white blood cells and platelets. Each of these carries out a different function.
- Soluble substances are transported in plasma.

1:9 Heart beat

Pumping the system

Blood flows around your body through a network of blood vessels which make up your **circulatory system**. Blood is kept flowing around the system by the pumping action of your heart. Your heart is the circulatory system's pump.

Pumping at between 60-70 beats per minute, which is about 100 000 beats every day, your heart is working all the time without pausing to rest.

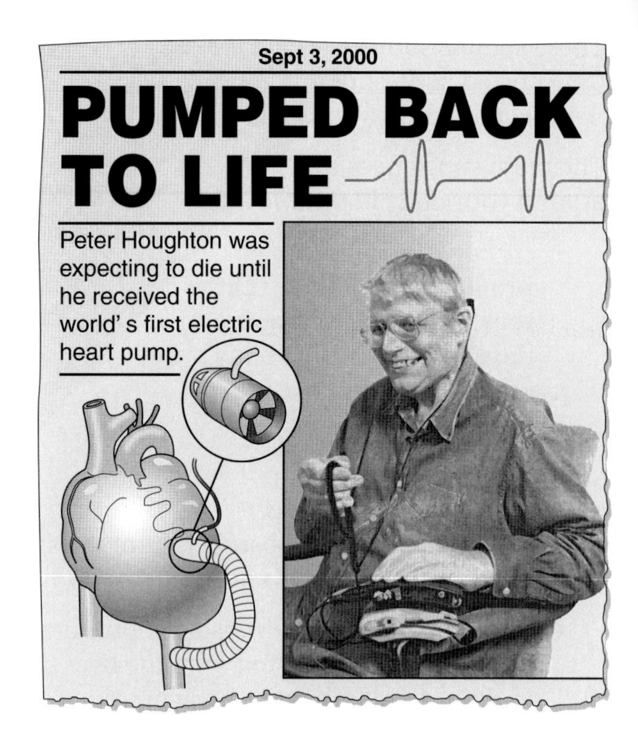
Sept 3, 2000

PUMPED BACK TO LIFE

Peter Houghton was expecting to die until he received the world' s first electric heart pump.

A double pump

If you follow the arrows in the diagram on the right you will trace the path of blood as it flows around the body.

Blood from the *right hand side* of the heart is pumped to the lungs where it picks up oxygen before returning back to the heart. Blood from the *left hand side* of the heart is pumped all around the rest of the body before returning to the heart.

This is called a **double circulatory system** because blood travels through the heart twice as it flows around the body.

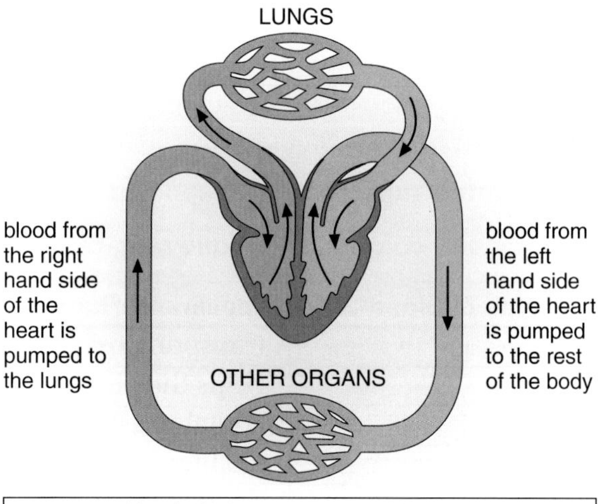

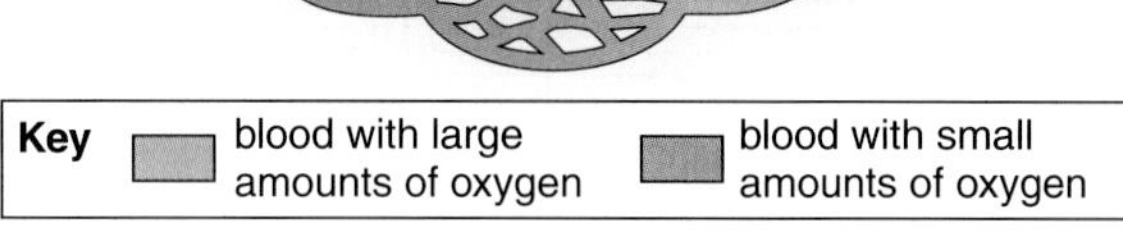

A pump with four chambers

The diagram below shows what the inside of your heart looks like. You can see that it is divided into four chambers – two **atria** and two **ventricles**.

Blood flows to and from the heart through the blood vessels connected to it.

Blood flows *from* the heart through **arteries**. Blood flows *back to* the heart through **veins.**

There are valves inside the heart and in blood vessels to ensure that blood flows in the right direction.

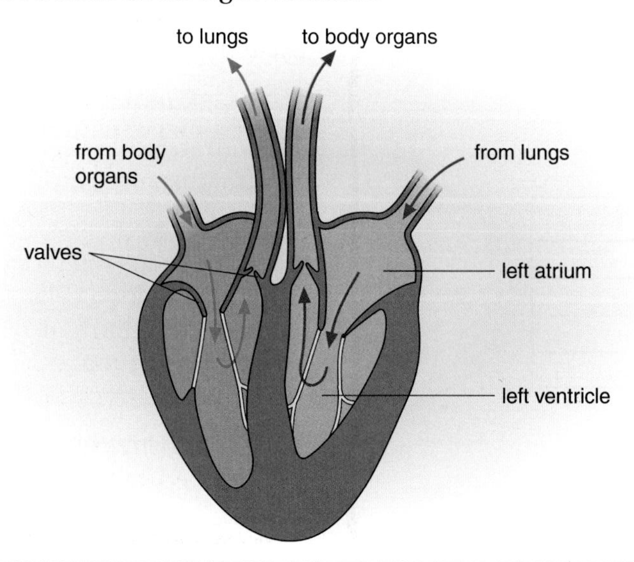

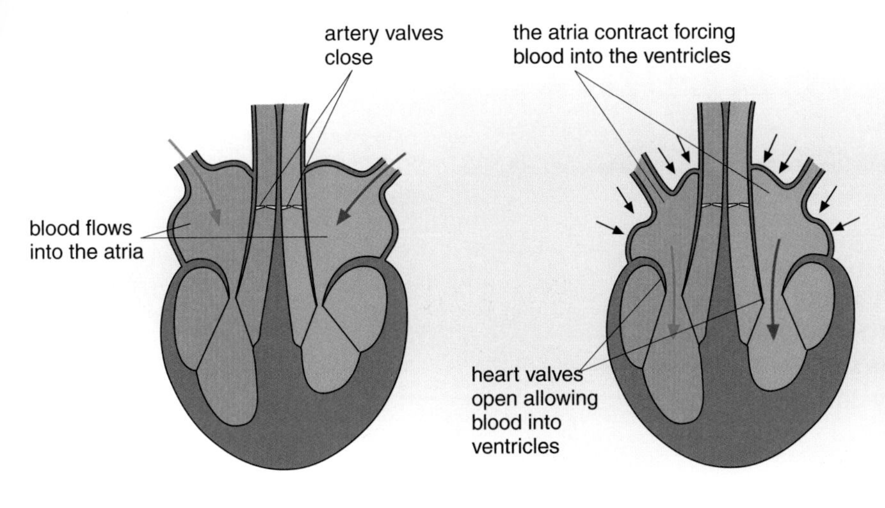

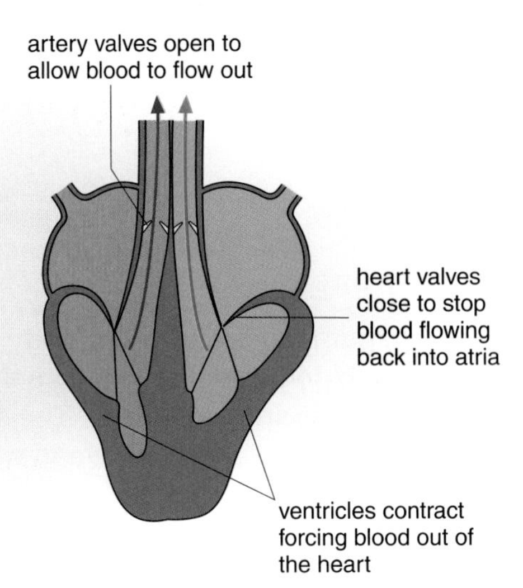

Your heart muscles and valves ensure blood flows in the right direction.

Heart action

The pumping action of your heart is produced by its thick muscular walls. When the muscles around the chambers *contract* blood is squeezed out. When the muscles *relax* blood enters the chambers. The heart valves and the artery valves only allow blood to flow in one direction.

Measuring heart action

Each time your heart beats blood is forced into arteries. When this happens the arteries expand as blood flows through them. Once the blood has passed through the arteries contract. This is your **pulse**. A pulse in an artery happens once every heartbeat. The table opposite shows the results of measuring pulse rates in a class of students.

Pulse rate (per min)	Pulse rate (per min)
80	70
70	74
70	77
72	72
75	70
68	80
70	71
72	74
73	72
71	69

Pulse rates in a class of 20 students.

Questions

1 a Explain what a pulse is.

b Copy and then use the results in the table above to complete this table:

Pulse rate (per minute)	Number of students
69	
70	
71	
…	

c Calculate the average pulse rate in this class.

d Measure your own pulse rate in beats per minute. Repeat this measurement every five minutes over a 30 minute period. Produce a table of your results.

e Explain why your pulse rate varies.

2 Write the list below in the correct order of circulation starting from blood flowing from body organs.

blood flowing from body organs **left atrium**
right atrium **right ventricle** **left ventricle**
blood flowing to body organs **blood flowing to lungs**

Summary

- The heart pumps blood around the circulatory system.
- There are two separate circulatory systems, one to the lungs and the other to all other body organs.
- The heart has four chambers.
- Valves in the heart ensure that blood flows through it in the right direction.
- Blood flows into the heart in veins and out in arteries.

1:10 Blood transport

Supply lines

Blood is pumped around your body through a highly complex system of blood vessels. There are three types of blood vessel – **arteries**, **veins** and **capillaries**. The diagrams show what each type of blood vessel looks like and what they do.

Arteries

Blood flows from your heart to all the organs in your body through arteries.

The blood in arteries is under high pressure because blood is pumped into arteries by the heart. Arteries have thick walls containing muscle and elastic fibres to withstand the pressure.

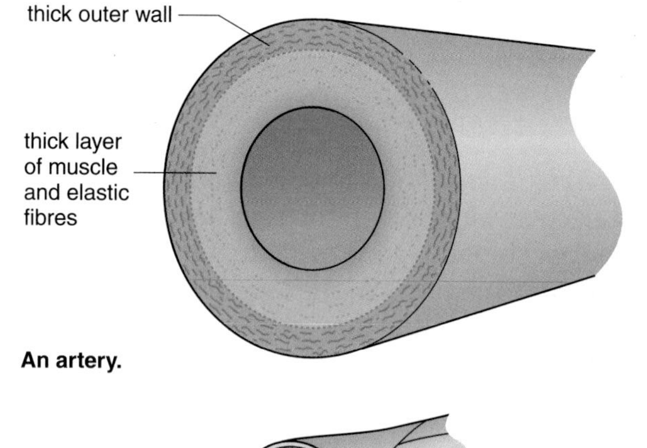

An artery.

Capillaries

Arteries branch to form smaller and smaller blood vessels and eventually form capillaries. Capillaries are found inside organs. They are very narrow and have thin walls which are only one cell thick.

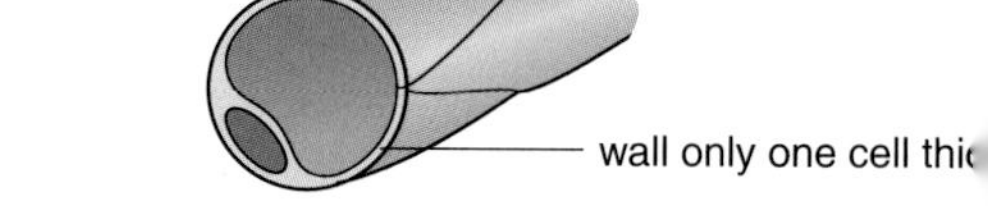

A capillary.

Veins

Capillaries gradually join up again to form veins. The pressure in veins is very low so blood flows much more slowly than in arteries.

Veins have thin walls as they do not have to withstand high pressure. They also have valves to stop any blood flowing backwards.

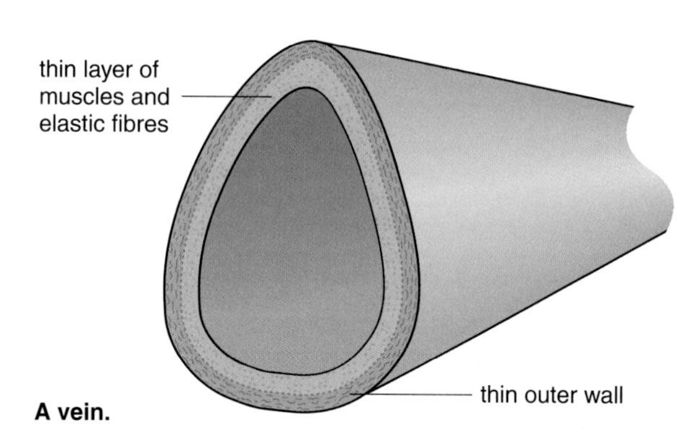

A vein.

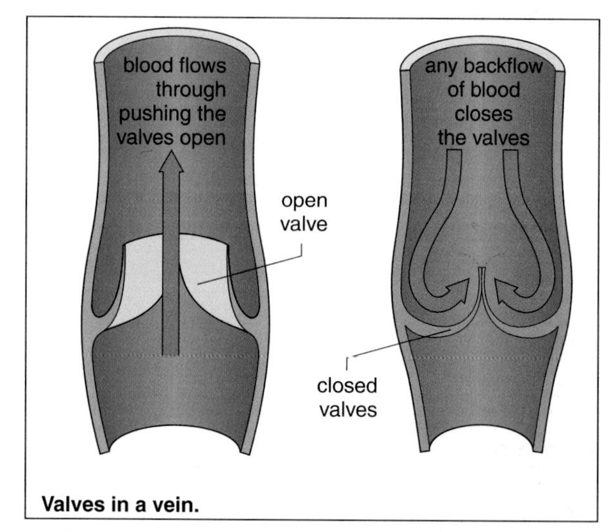

Valves in a vein.

a **Blood flows away from the heart pump in arteries. Explain:**

- i **Why the blood pressure in arteries is high.**
- ii **Why arteries have thick walls.**
- iii **Why arteries do not have valves.**

Exchanging substances

The thin walls of capillaries allow substances needed by cells to pass *from* the blood. The thin walls also allow waste substances produced by cells to pass *into* the blood.

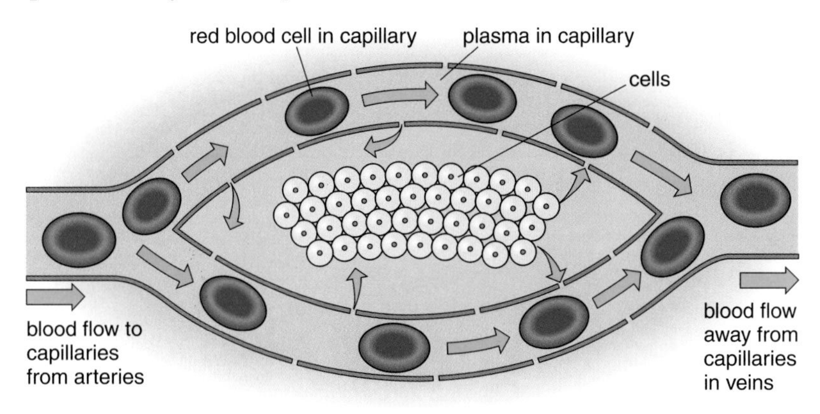

Moving in and out

Capillaries are the part of your circulatory system where substances enter and leave the blood. Below are three examples of organs where substances move in and out of blood through capillaries.

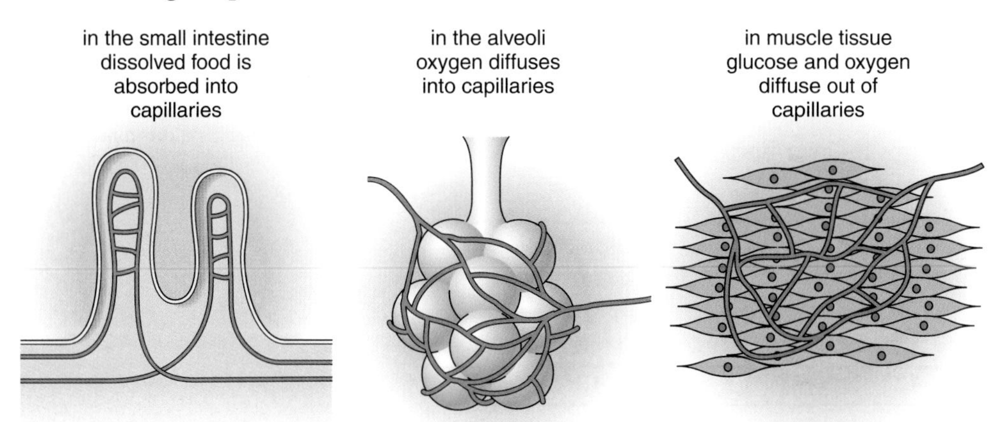

Substances move in and out of capillaries.

Questions

1 Copy and complete the following sentences using the words provided:

muscle fibres veins capillaries

Blood is carried back to the heart in __________. Arteries have thick walls made from __________ and __________. The smallest blood vessels are called __________ and have walls

Copy and complete this table:

	Veins	Arteries	Capillaries
wall thickness (thick or thin)			
valves (present or absent)			
blood pressure (high or low)			

Summary

- Arteries, veins and capillaries have different structures.
- Substances are exchanged between cells and the blood through capillary walls.
- Arteries have thick walls to withstand high pressure.
- Veins have valves to stop backflow of blood.

1:11 Making the body work hard

Britain is getting fatter

The health of most people living in Britain is now better than it used to be. One aspect of health that has become worse is the increase in the number of adults who are overweight. Britain is becoming a fatter nation.

Being overweight increases the risk of heart disease, high blood pressure and other illnesses.

Regular exercise helps to stop people becoming overweight. Despite the benefits of exercise nearly 20% of British adults do no or very little exercise. Activity levels of British youngsters still at school are very low. For example, half of primary school aged children do not take enough exercise.

Taking part in sport can be fun as well as improving health.

Getting enough energy

When you are exercising your muscles need more energy. This energy is released from glucose during respiration. When muscles are working harder more respiration takes place to release more energy.

The table below shows some of the changes in your body as you work harder.

	Breathing rate (number of breaths per min)	Pulse rate (number of pulses per min)
at rest	18	65
after 10 minutes exercise	42	130

From the table:

a Draw a bar chart to show pulse rate at rest and during exercise.

b Compare the resting pulse rate with pulse rate after exercise.

c Explain why pulse rate and breathing rate both change during exercise.

Improving fitness

Getting involved in regular exercise improves your fitness. This is because exercise has several effects on your body. When you exercise regularly your:

- heart muscles become stronger so that the heart pumps blood round your body more efficiently;
- rib muscles and diaphragm become stronger making your breathing system more efficient;
- muscles become larger and able to work hard for longer.

Pulse rates

Measuring pulse rate before and after exercise can show how fit you are. In a fit person the pulse rate is often low. When a fit person exercises his or her pulse rate increases and then it quickly returns to normal. When an unfit person exercises his or her pulse rate may go very high and it returns to normal only slowly.

Questions

The graph opposite shows the pulse rates of two girls. One of them takes part in sport and the other takes no exercise at all.

1 From the graph:

a Which of the two girls is the fitter?

b What evidence is there on the graph to support your answer?

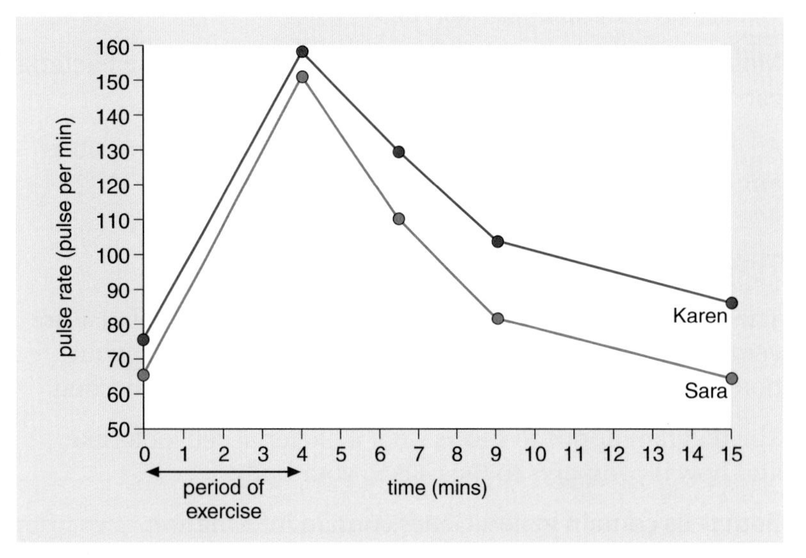

2 The effects of regular exercise on your body can be seen in the table below.

	Before regular exercise	After regular exercise
pulse rate (number of beats per min)	72	63
amount of blood pumped out of the heart during each beat (cm^3)	120	140
breathing rate (number of breaths per min)	15	12

The effect of taking regular exercise. All measurements are taken when the person is at rest.

Using the table:

a Describe how pulse rate changed as the person became fitter.

b What evidence is there to show that regular exercise can make

i Heart muscles stronger?
ii Rib muscles and diaphragm stronger?

Summary

- When you are active your body uses more energy.
- To provide more energy the rate of respiration in cells increases.
- To increase their rate of respiration cells need more glucose and oxygen.
- To provide cells with more glucose and oxygen the heart beats faster and stronger and breathing becomes faster and deeper.

1:12 Microscopic enemies

Harmful microbes

Very small organisms are called **microbes**. **Bacteria** and **viruses** are examples of microbes.

Most bacteria are harmless and some are useful. Some bacteria can cause disease and make you ill.

All viruses are harmful. They cause diseases in humans, other animals and in plants.

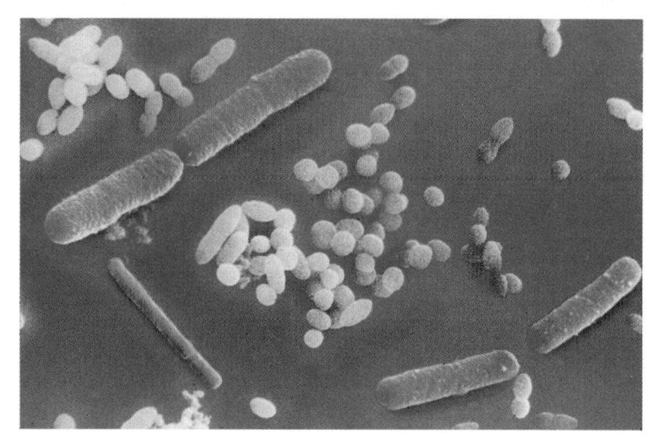

Bacteria are tiny single-celled organisms.

Tiny invaders

The bacteria which can enter your body and cause disease are very, very small. The cells of bacteria are smaller than your body cells and can only be seen with a powerful microscope.

The diagram opposite shows what a bacterial cell looks like and how it compares to the cells in your body.

Both cells contain **genes**. Genes contain information which controls what happens in a cell. The genes are found in **chromosomes**. In human cells chromosomes are found inside a nucleus. The chromosomes of bacteria are not in a nucleus.

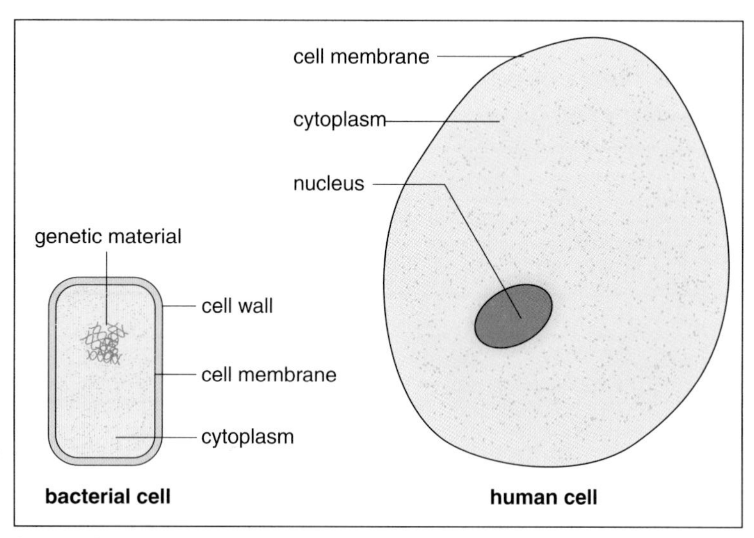

A bacterium and a human cell magnified 2000 times.

a List two features which are found:

i in both human cells and bacterial cells
ii in the bacterial cell and not in the human cell.

Rapid reproduction

Bacteria can reproduce very quickly. Bacteria reproduce by dividing into two. If the conditions are right bacteria may divide every 20 minutes. This means that a small number of bacteria can increase to thousands in a very short time.

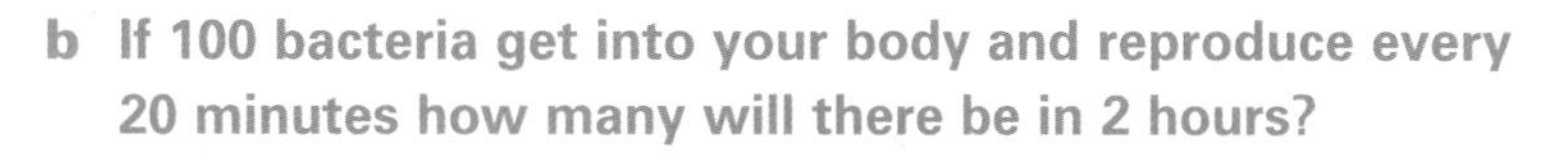
b If 100 bacteria get into your body and reproduce every 20 minutes how many will there be in 2 hours?

You can work this out by copying and completing the table:

Time (min)	Number of bacteria
0	100
20	
40	
60	
80	
100	

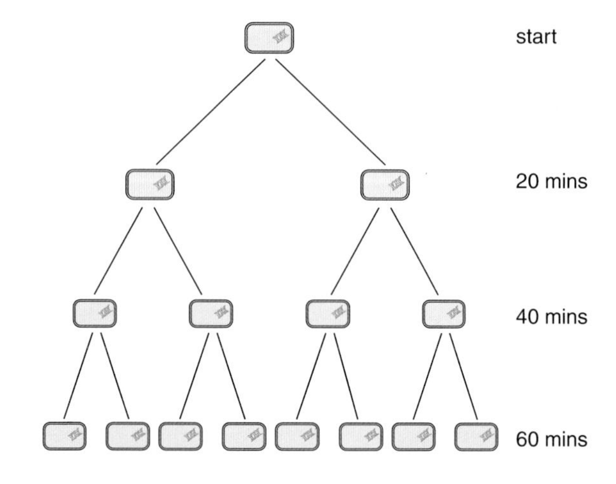

The number of bacteria doubles after each division.

The smallest invaders

Viruses are even smaller than bacteria. The diagram opposite shows what viruses are like. They have a coat made of protein. This coat surrounds a few genes.

Viruses cannot reproduce by themselves. They can only reproduce inside living cells. Viruses invade living cells, such as human cells. When a virus gets into a living cell it uses it to make thousands of new viruses. The new viruses burst out of the cell and invade other cells. This damages the cell.

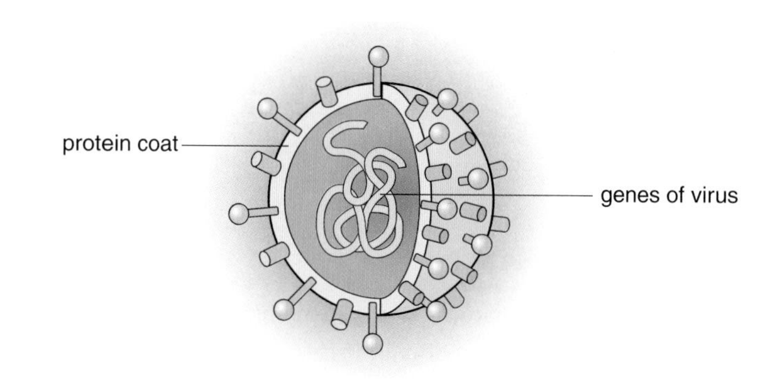

A virus magnified many times.

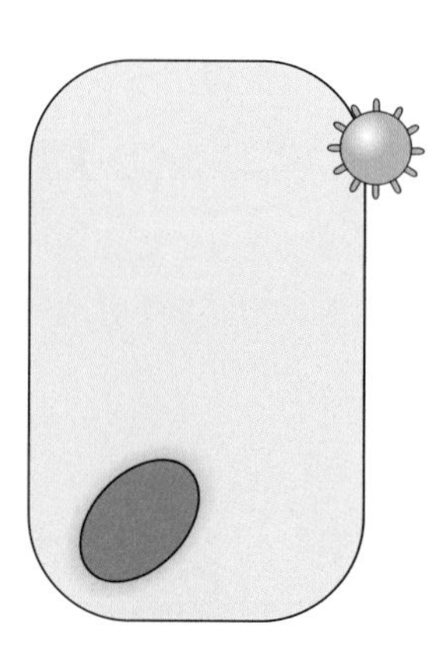

the virus attacks a cell...

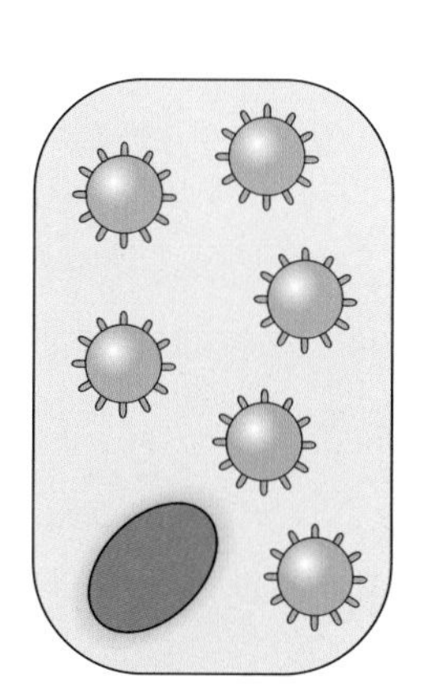

...enters the cell and new viruses are made inside the cell...

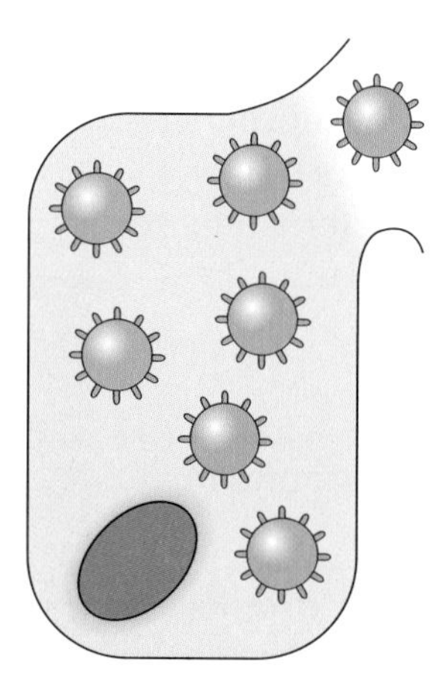

...then the cell bursts, releasing new viruses to infect more cells

How a virus reproduces.

Questions

1 Copy and complete the sentences below using words from this list.

genes nucleus chromosomes wall respire reproduce protein

The cells of bacteria contain cytoplasm, cell membranes, and cell __________. Their chromosomes are not in a __________. Viruses can only __________ inside living cells. Viruses have a __________ coat surrounding a few __________.

2 How do bacteria reproduce?

3 List all the differences you can find between bacteria and viruses.

4 How does a virus damage cells?

Summary

- Viruses and bacteria are examples of microbes.
- A bacterial cell has cytoplasm and a membrane surrounded by a cell wall.
- Unlike the genes in human cells the genes in a bacterial cell are not found in a distinct nucleus.
- Viruses have a protein coat surrounding a few genes.
- Bacteria and viruses can reproduce very quickly inside your body.
- Viruses can only reproduce after they have got inside a living cell.
- Viruses damage the cells in which they reproduce.

1:13 Spreading diseases

Catching a disease

Bacteria and viruses are types of microbe. Many diseases are caused when microbes get into the body.

You are more likely to develop a disease if large numbers of viruses or bacteria enter your body. This can happen if you are exposed to unhygienic conditions or by being in contact with someone who already has a disease.

Once viruses or bacteria are in your body they reproduce rapidly. There are many places inside you where bacteria and viruses can live and reproduce. Some bacteria release poisons called **toxins** which make you feel ill.

Would you buy your hotdog here?

Spreading harmful microbes

Microbes which cause disease can be spread around in many ways. Here are just a few examples.

When people cough or sneeze they release tiny droplets of moisture into the air. Microbes can stick on the droplets and can get into your body when you breathe in. Common diseases such as colds and 'flu are spread in this way.

Food poisoning is one of the most common illnesses. You become ill by eating food containing certain bacteria. This is why it is important to store and handle food in hygienic conditions.

Insects and other animals such as rats can spread harmful microbes. Houseflies spread bacteria from one place to another as you can see in this poster.

Coughs and sneezes spread diseases.

It is important to keep food in hygienic conditions.

This is what happens when a fly lands on your food.
Flies can't eat solid food, so to soften it up they vomit on it.
Then they stamp the vomit in until it's a liquid, usually stamping in a few germs for good measure.
Then when it's good and runny they suck it all back again, probably dropping some excrement at the same time.
And then, when they've finished eating, it's your turn.

Cover food. Cover eating and drinking utensils. Cover dustbins

Health Education Authority

Major disasters

Major disasters, such as the floods which happened in Mozambique in February 2000, can kill hundreds of people.

People still faced danger even when the floods began to disappear. This is because diseases such as cholera spread quickly following disasters. Cholera is caused by bacteria which are spread in water that has been contaminated with human sewage. In developing countries a toilet may be a deep hole dug into the ground. When the ground floods bacteria in the sewage are carried in the floodwater.

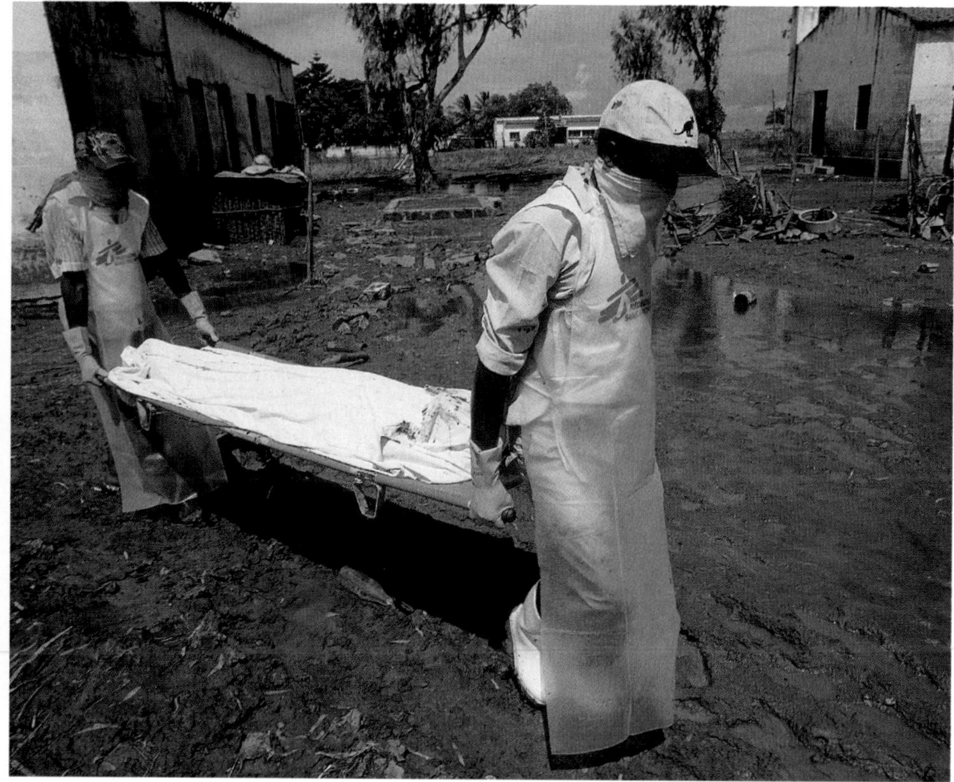

Rescue workers clearing up after floods must wear protective clothing to avoid infection.

Questions

1 Explain why being on a crowded bus with people who have colds can increase your chances of developing a cold.

2 List three ways that shops handling food such as meat prevent the food becoming infected with bacteria.

3 Copy and complete the following sentences using the words provided:

contaminated	**microbes**	**animals**
bacteria	**moisture**	**disasters**

Cholera is a disease caused by __________. People can develop this disease by drinking __________ water. 'Flu is a common disease caused by __________ which is spread on __________ in the air.

4 Explain why it is important to cover food to keep houseflies away.

Summary

- You are more likely to develop a disease if large numbers of microbes enter your body.
- Microbes reproduce when they get inside the body.
- Some microbes release toxins.
- Microbes can easily spread.

1:14 Body defences

Outer protection

Your body is always coming into contact with harmful microbes without you becoming ill. This is because your body is very good at keeping microbes out.

Your **skin** is a barrier stopping microbes getting inside your body.

If your skin is cut microbes may get in and cause infection. To stop this happening your blood forms **clots** that seal any cuts. The clotted blood is the scab that forms over a cut. The clot prevents entry of harmful microbes and stops you losing too much blood.

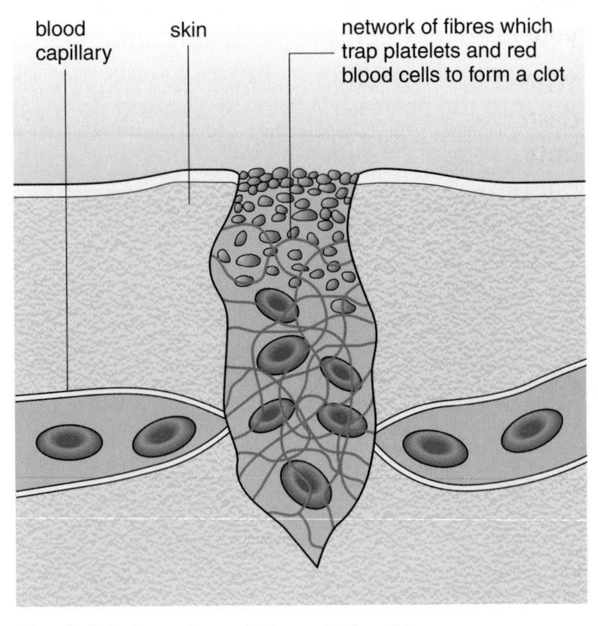

Blood clots to seal a cut through the skin.

Trapping microbes

Harmful microbes can enter your body in the air you breathe.

Your breathing passages are lined with cells which produce a sticky liquid called **mucus**. Bacteria that are breathed in get stuck in the mucus.

There are other cells lining your breathing passages which have tiny hair-like projections called **cilia**. The movement of the cilia sweeps any trapped bacteria back up to your throat where they can be swallowed.

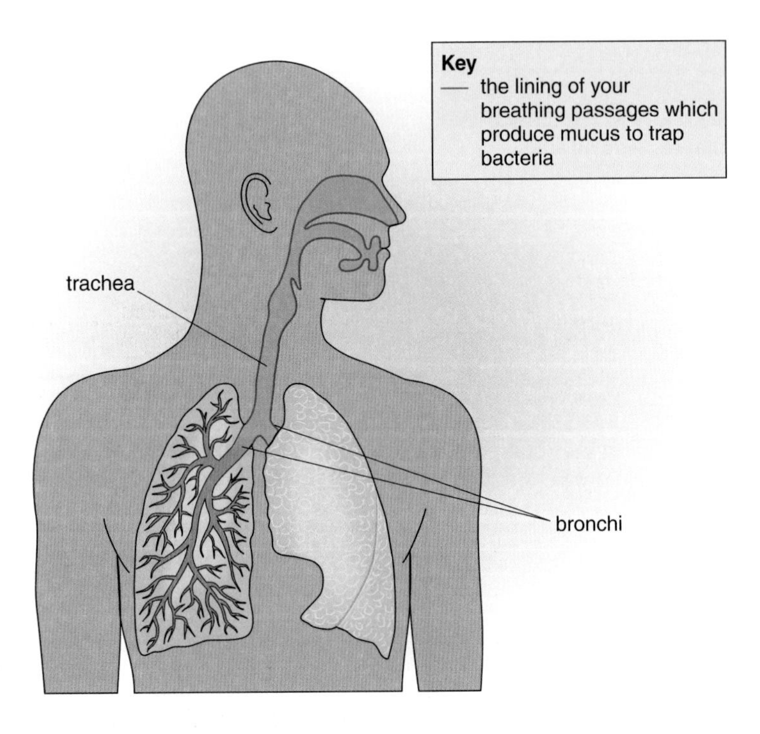

a List three ways by which harmful microbes are stopped from getting into your body.

Cells to fight diseases

If microbes do get into your body you are protected by your **white blood cells**. These cells are a very important part of your body's defence system against microbes.

One way that your white blood cells protect you against disease is by taking in and digesting microbes which get into your body.

White blood cells release chemicals called **antibodies** which destroy microbes. They also release chemicals called **antitoxins** which prevent the toxins made by microbes from poisoning your body.

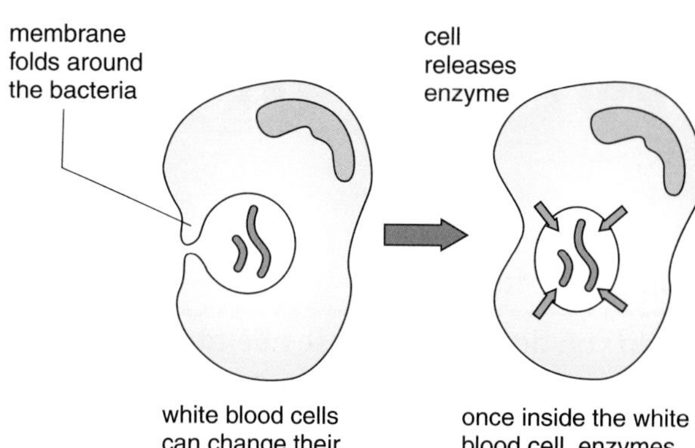

white blood cells can change their shape and wrap around bacteria

once inside the white blood cell, enzymes are released to digest the bacteria

microbes come into contact with white blood cell

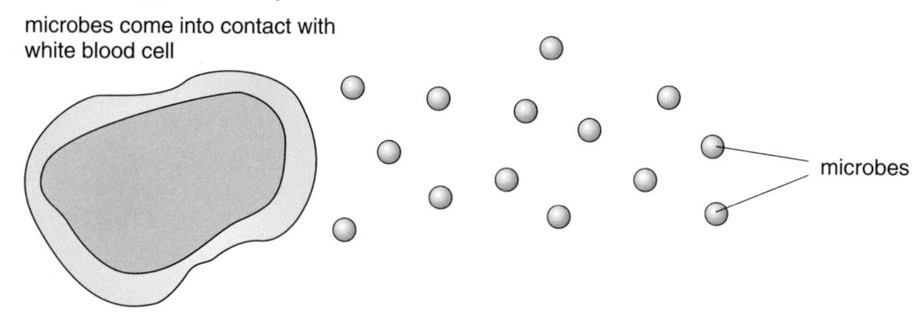

white blood cell releases **antibodies**

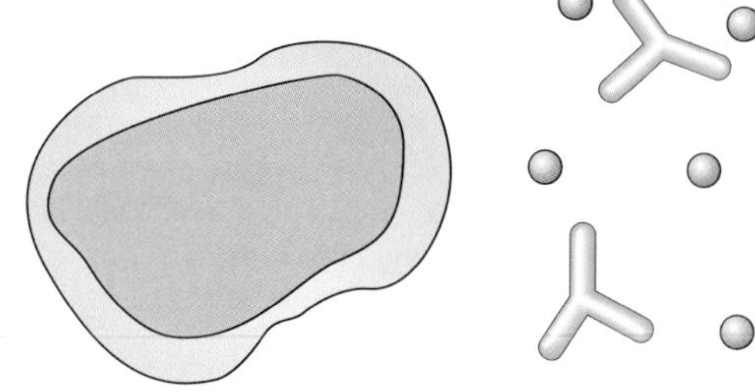
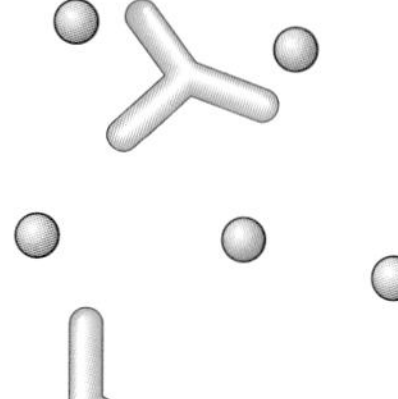
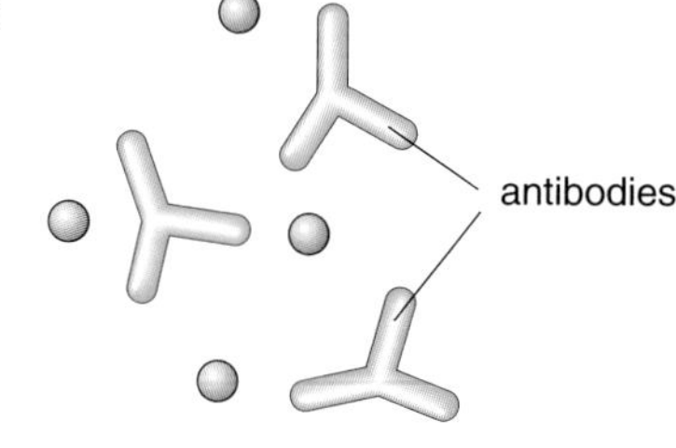

antibodies react with microbes and destroy them

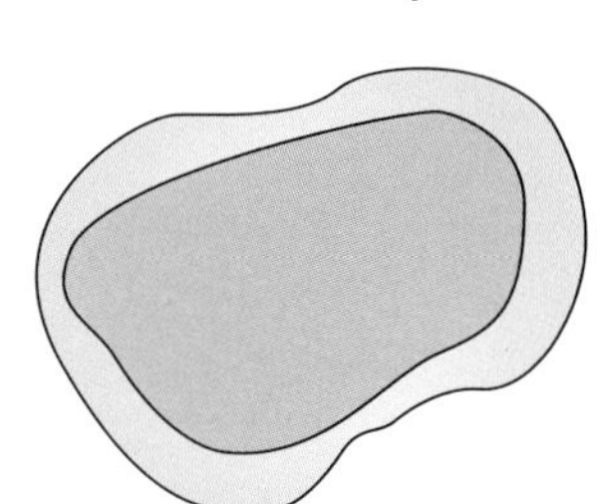
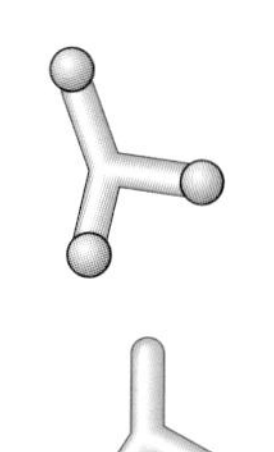
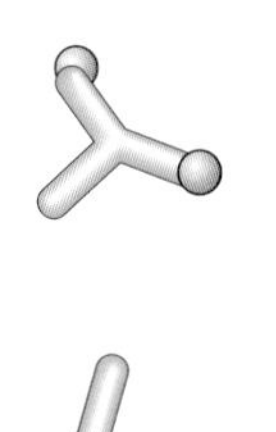
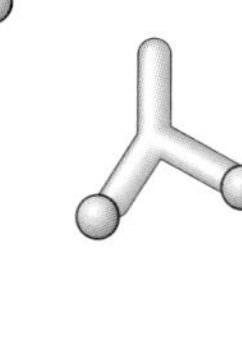

Questions

1 Copy and complete the following sentences using the words provided:

antibodies digest shape antitoxins cells

Some white blood cells take microbes into their cells and __________ them. Other white blood cells destroy microbes by releasing __________. They can also release __________ which prevent the toxins produced by microbes from working.

2 List the ways in which your body stops harmful microbes from getting inside.

3 What is the job of your white blood cells?

Summary

- Microbes are kept out of your body by your skin, blood clots and the mucus produced by breathing passages.
- White blood cells destroy micobes by digesting them and by producing antibodies.
- White blood cells also release antitoxins.

1:15 Life-long protection

Immunity for life

When you develop a disease microbes will be reproducing inside your body. Fortunately, your white blood cells will be destroying the microbes, helping you to recover.

Once your white blood cells have destroyed a type of microbe you are unlikely to develop the same disease again. This is because your white blood cells can recognise the type of microbe if it gets into your body again. Your white blood cells then produce antibodies much more quickly. The microbe will no longer be able to make you ill. This makes you **immune** to the disease for the rest of your life.

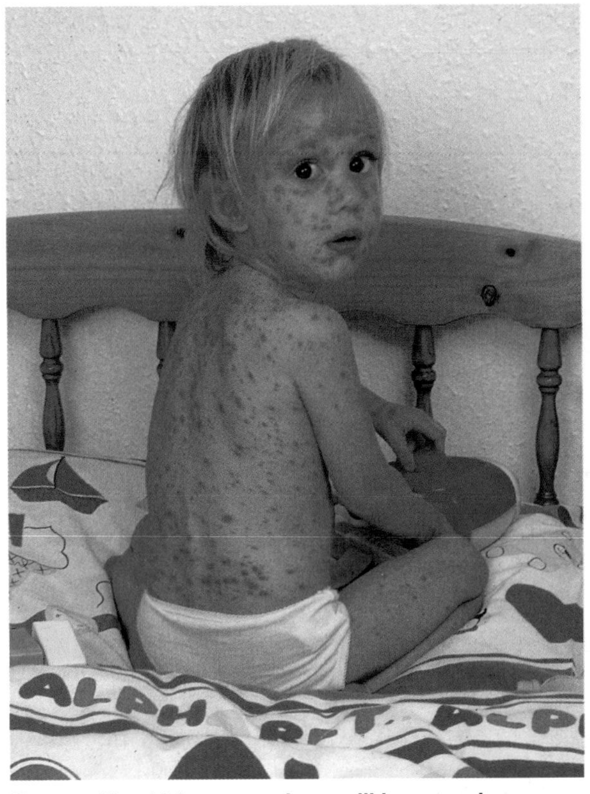

Illnesses like chickenpox make you ill because viruses damage your cells. Your white blood cells help you to recover.

What happens when you develop a disease?

When microbes get into your body you may develop a disease.

Greg was away from school with 'flu. The graph below shows what happened in his body after he became infected with the 'flu virus.

a For how long did Greg feel ill with 'flu?

b How long did it take viruses to start to reproduce in Greg's body after he was first infected?

c On which day did the number of viruses start to decrease?

d What made the number of viruses in Greg's body decrease?

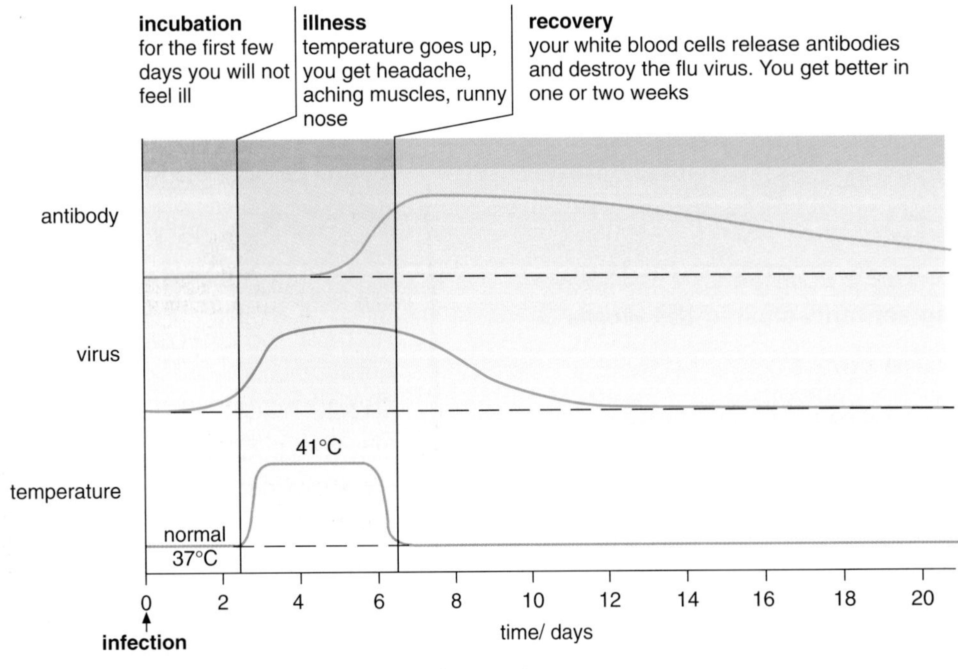

Changes which take place when someone develops 'flu.

A quick jab

When you were a young child you were probably **vaccinated** against a number of diseases, such as whooping cough, measles and polio.

Vaccinations are given to protect you against certain harmful diseases. For example, you may have been recently vaccinated against TB (tuberculosis). This vaccination protects you against the bacteria which cause TB – a disease which can badly damage your lungs even after you have recovered from the disease.

Vaccinations involve injecting a weakened form of the microbe that causes the disease into your blood. Because the microbes are weakened, or even dead, you do not become ill but your white blood cells produce the antibodies that will give you immunity to the disease.

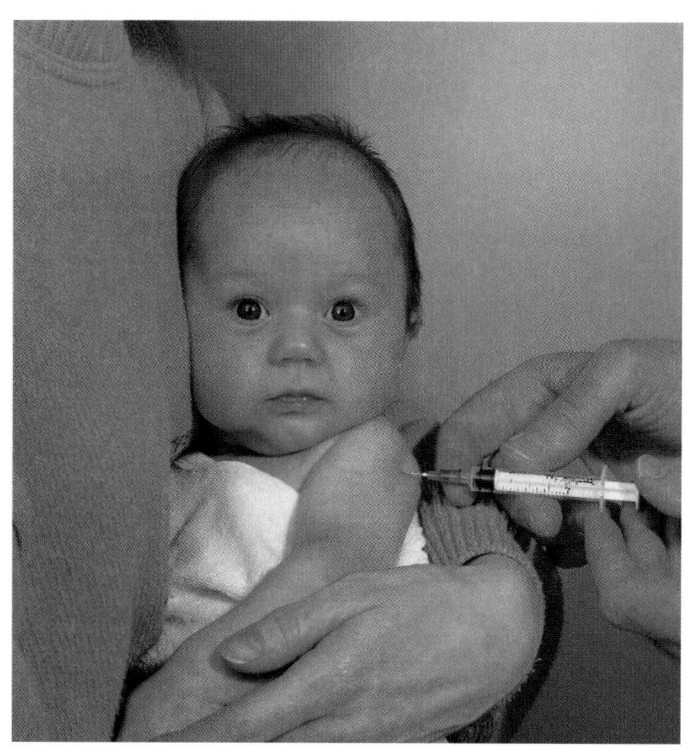

Immunisation when you are young protects you for life.

Dangerous diseases

Diseases such as TB, whooping cough and measles can be very harmful. This is why vaccines have been developed to give you immunity.

You have probably been given the MMR vaccine which protects you from Mumps, Measles and Rubella (German measles). Before this vaccine was available about 90 children a year in the UK died from measles. Because of immunisation children in the UK no longer die of measles.

The bar chart shows the effect of the immunisation programme against measles.

	1989	1996	1997	1998	1999
% of children receiving MMR vaccine	none	92	92	91	88
Number of cases of measles	24400	5300	3600	3500	2300

Question

1 Explain what the word 'immune' means.

2 What type of microbe gives you 'flu?

3 List diseases on this page that can be prevented by vaccination.

4 What diseases will the MMR vaccine protect you against?

5 **a** What percentage (%) of children were immunised with MMR in
 i 1996?
 ii 1998?

b MMR vaccine was first used in 1989. Describe how the number of people developing measles changed after MMR was in use.

Summary

- People can be immunised against disease.
- Vaccines contain mild or dead microbes.
- White blood cells respond to vaccines by producing antibodies.

End of module questions

1 The diagram shows a bacterial cell. Match words from the list with each of the labels1–4 in the diagram.

cell wall
cytoplasm
membrane
genetic material

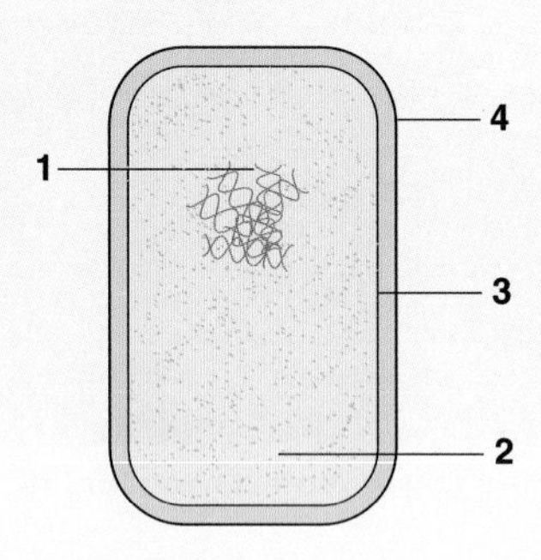

2 The table shows the jobs of different types of blood cells. Match the words from the list with each of the numbers in the table.

antibodies
oxygen
clots
antitoxins

Type of cell	What the cell does
red blood cell	carries __1__ around the body
white blood cell	produces __2__ to defend the body against disease
white blood cell	produces __3__ to overcome poisons
platelets	help to form __4__ to seal wounds

3 The diagram shows some of the structures in the heart. Match words from the list with each of the labels in the diagram.

contracts to force blood out of the heart
stops blood flowing back into the heart
contains muscle cells
transports blood to the lungs

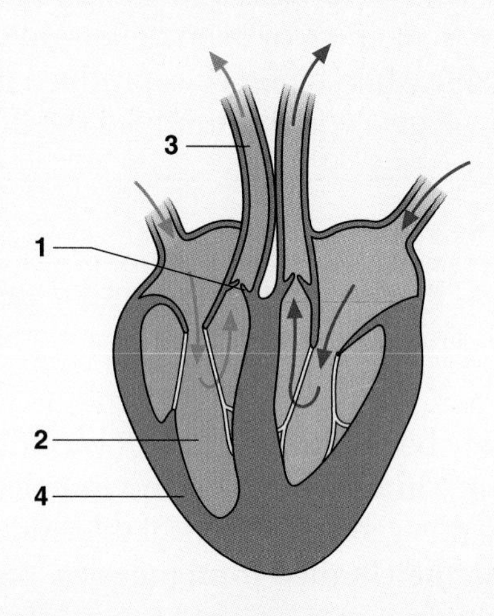

4 The diagram shows part of the lungs with a blood vessel showing the direction that blood flows.

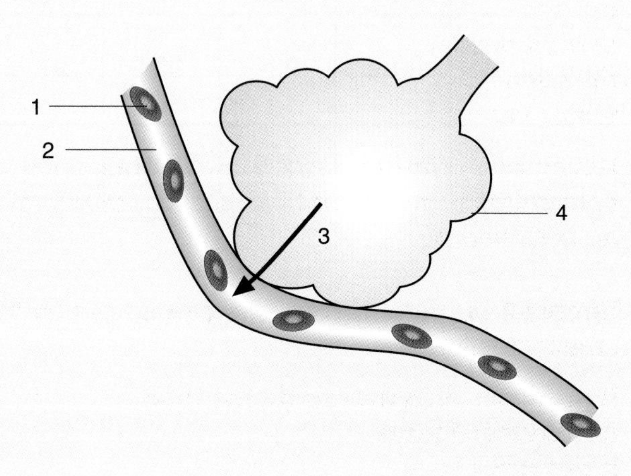

Match words from the list with each of the labels 1–4 on the diagram.

blood containing a high concentration of carbon dioxide
membrane which provides a large surface area for the diffusion of gases
diffusion of oxygen
wall of a capillary

5 The table shows the jobs of some of the substances produced by the digestive system.

Match words from the list with each of the numbers 1–4 in the table.

sugar
protease
lipase
bile

Substance	The job of the substance
carbohydrase	Breaks down starch into 1
2	Digests proteins to form amino acids
3	Digests fats to form fatty acids and glycerol
4	Forms tiny fat droplets from fats

6 Your breathing system is made up of several parts.

The movement of which **two** parts of the breathing system make air move into the lungs?

alveoli
diaphragm
ribcage
trachea
bronchioles

7 There are four different gases in the air breathed in and out. These are:

oxygen
carbon dioxide
nitrogen
water vapour

Which **two** of these gases are in higher concentration in the air breathed out than in the air breathed in?

8 Choose from the list **two** jobs carried out by the small intestine.

it produces enzymes to digest food
it produces faeces
it produces bile
it has a large surface area to absorb food

9 The digestive system produces enzymes to break down large food molecules.

The diagram shows part of the digestive system.

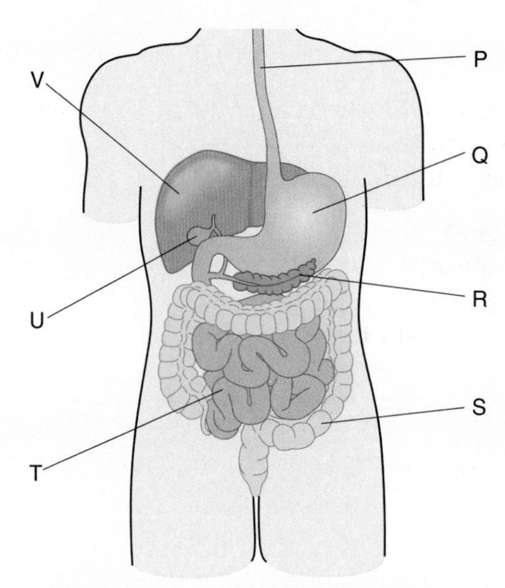

9.1 Which part is the pancreas?

A P
B Q
C T
D U

9.2 Most of the products of digestion are absorbed at?

A S
B T
C U
D V

9.3 In part labelled U.

A starch is absorbed
B enzymes work in alkaline conditions
C protease enzymes digest proteins
D bile is released to emulsify fats

9.3 Lipase enzymes are produced by

A R and T
B R and U
C V and U
D T and Q

10 The diagram shows an investigation to find out how food is absorbed from the digestive system.

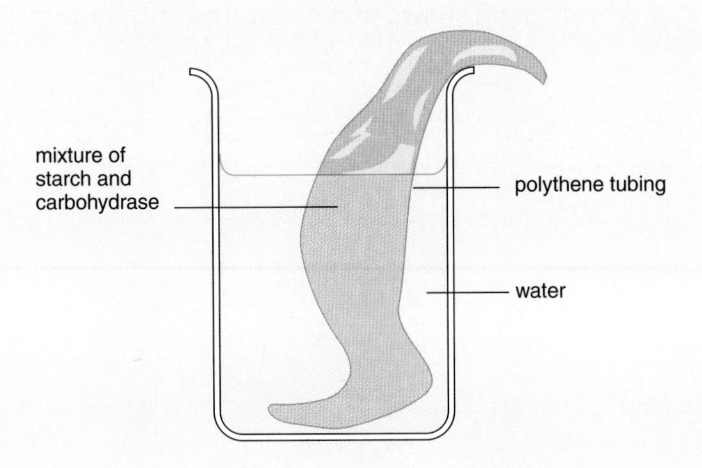

10.1 The polythene tubing used in this investigation works in the same way as the

A alveolar surface
B wall of the small intestine
C wall of capillaries
D cell membrane

10.2 After about 30 minutes the water around the tubing will contain

A starch
B carbohydrase
C protein
D sugar

10.3 Substances can pass through the tubing by

A diffusion
B pressure
C filtration
D contraction

10.4 Substances which pass through the tubing are

A large
B insoluble
C enzymes
D small

11 The graph below shows the pulse rates of two girls measured before, during and after taking the same amount of exercise.

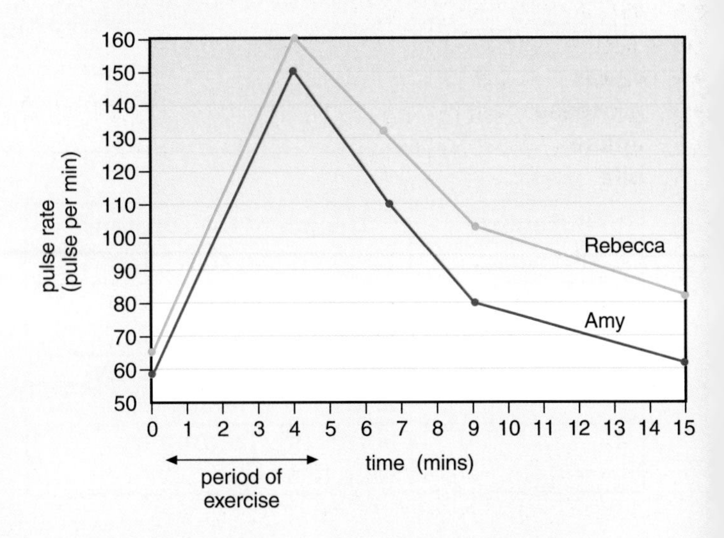

11.1 The pulse is measured by feeling a

A muscle
B capillary
C artery
D vein

11.2 Evidence that Amy is the fitter of the two girls is

A she has a lower pulse rate at rest
B her pulse rate gets back to normal more quickly
C her pulse rate goes higher during exercise
D her pulse rate gets back to normal more slowly

11.3 Pulse rate increases because

A the heart beats more rapidly
B each heart beat is stronger
C blood pressure is higher
D breathing rate increases

11.4 The increase in pulse rate increases the supply of

A lactic acid to muscles
B energy to muscles
C heat to muscles
D sugar to muscles

Module 2 – Maintenance of life

The world can be a difficult place for life to survive in. For plants and animals it is important to get the energy they need to live and grow from their surroundings. In this unit you will find out about how plants and animals maintain their life processes and some of the difficulties they might face.

Humans and animals use their sense organs to detect changes in the world around them and respond to them. This allows them to survive and take what they need from the environment. Humans and most animals can use their brains to make decisions.

Plants, on the other hand, must get everything they need to survive from the air and soil around them. They need to be able to detect light and water in order to grow well.

In this unit you will study some important life processes and find out how these help living things survive.

Try these first

1. What are the main differences between plant and animal cells?
2. What are the main sense organs in the human body?
3. How do people get the energy they need to survive?
4. What is the process plants use to make their food?
5. Why do humans and animals need to breathe?

2:1 Detecting changes

The world around us is always changing. Animals need to detect these changes in order to survive.

Receptors

The photograph shows an animal called the loris. The loris lives in tropical forests. It sleeps during the day, but at night it hunts the small insects that it feeds on.

a Why do you think the loris is a very successful night-hunter?

The loris detects its prey by sight. Sharks also need to detect their prey, but they have very poor eyesight. However, sharks are able to detect a wounded animal from a great distance.

b How do you think sharks detect wounded animals?

Impala are prey for large cats and other animals. They need to detect leopards to avoid being eaten.

c How do you think impala can detect leopards that are crawling through grass towards them at night?

The surroundings of an animal are called its **environment**. Animals use specialised cells called **receptors** to detect changes in the environment. A change in the environment that can be detected by a receptor is called a **stimulus**.

Humans have receptors that are sensitive to:

- light
- sound
- touch and pressure
- chemicals
- changes in temperature
- changes in position

These receptors are found in different parts of the body. A part of the body containing many receptors is called a **sense organ**. Some of the body's sense organs are shown in the diagram below.

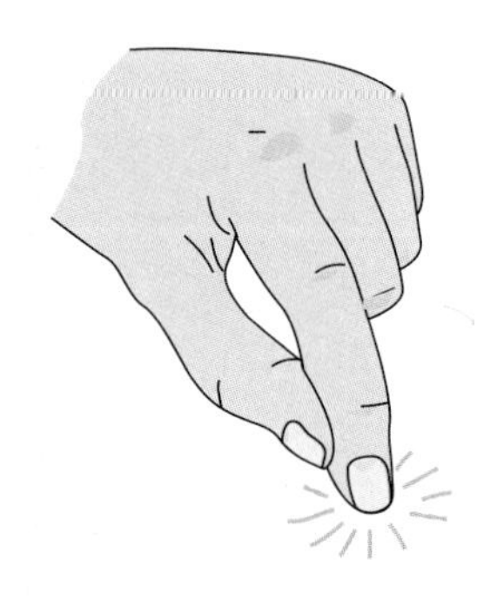
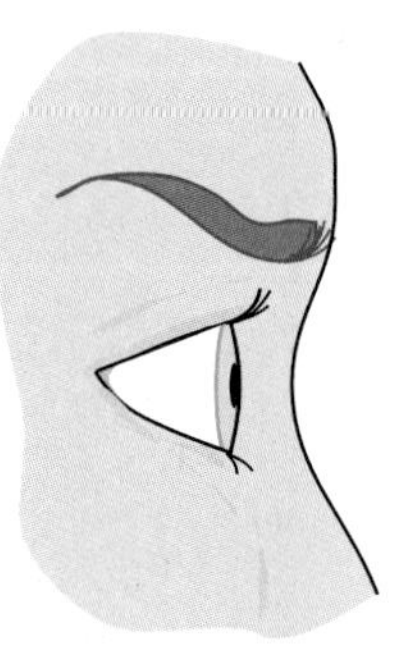
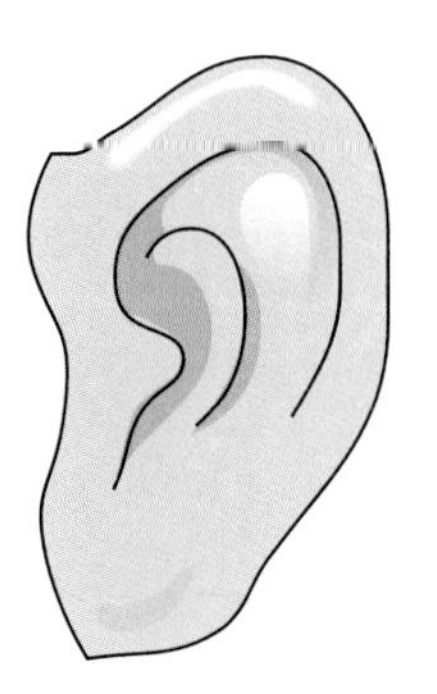
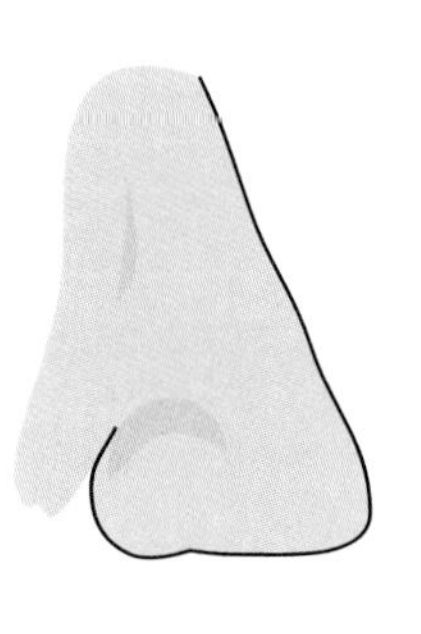
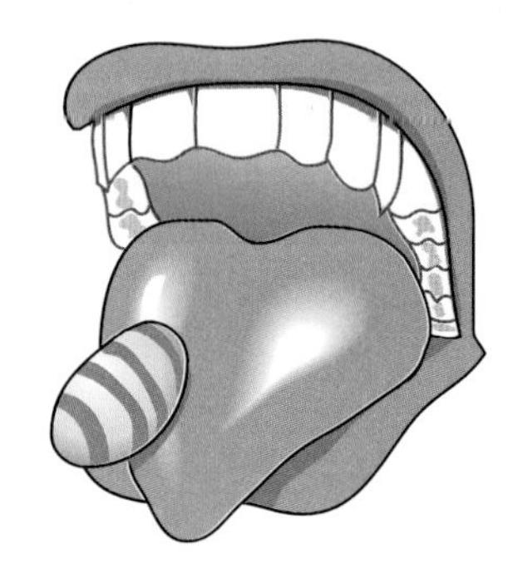

d Which of the sense organ(s) in the diagram on the previous page contains mainly receptors sensitive to:

i Sound?
ii Chemicals?
iii Light?
iv Touch and pressure?

If pressure receptors are over-stimulated then we feel pain. That is why a tap does not hurt but a smack does.

Reacting to stimuli

Receptors only detect **stimuli**. The information from the receptors needs to be passed to the **brain**.

Information is carried round the body by specialised cells called nerve cells (**neurones**). Nerve cells gather together to form **nerves**.

The brain and the spinal cord consist of countless millions of nerve cells. Nerve cells from parts of the body below the head enter or leave the spinal cord, rather than the brain itself. The spinal cord carries the information to or from the brain.

The brain and spinal cord collect information from receptors, make sense of it, and then send messages to the organs that need to respond. For example, if the touch receptors in your finger tell the spinal cord that they have touched something too hot, the spinal cord makes the muscles in the hand move away.

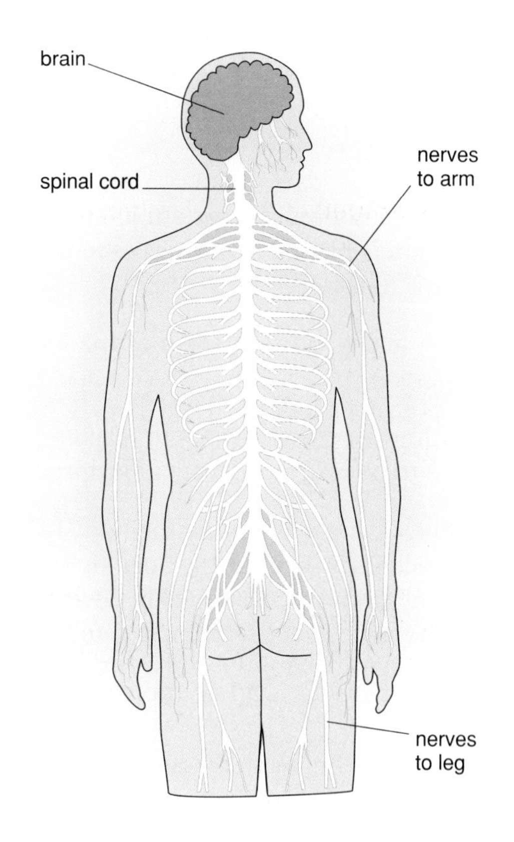

Information is carried around the body by specialised nerve cells.

Questions

1 Copy and complete the following sentences using the words provided:

stimulus receptors nerve cells spinal cord brain

A change in the environment is called a __________. Animals have __________ to detect these changes. Information is carried around the body by __________ __________. The nervous system consists of nerves, the __________ and the __________ __________.

2 List the main sense organs in the body.

3 In accidents, the spine is sometimes damaged. The person may then be paralysed below the damaged part of the spine. Explain why this happens.

Summary

- Cells called receptors detect stimuli. These include receptors in the:
  - eyes which are sensitive to light;
  - ears which are sensitive to sound;
  - tongue and in the nose which are sensitive to chemicals.
- The nervous system consists of nerve cells called neurones.
- Neurones are gathered together to form nerves.

2:2 Reflex actions

The brain does a very important job in the nervous system. Information is passed along nerves to the brain whenever a stimulus is detected.

The brain makes sense of the information and co-ordinates the response.

It takes time for the messages to reach the brain and a response to be returned to the appropriate organ.

Sometimes it is necessary for the body to react more quickly, for example if you are getting burnt it is important for you to move away from the heat quickly. Severe pain detected by receptors may cause a reflex action to take place.

The knee-jerk reflex is an example of a **reflex action**.

The knee-jerk reflex

The nurse is testing the patient's nervous system. She taps the tendon just below the knee with the rubber hammer.

If the knee jerks upwards automatically, the nervous system in the leg is working correctly. The diagram shows how the nervous system makes this happen.

The patient's leg responds very quickly. The response of the leg is automatic. The patient does not need to think about moving their leg. Responses that we do not have to think about need not involve the brain. Automatic reactions like these are called reflex actions. We are born with reflex actions – we do not learn them.

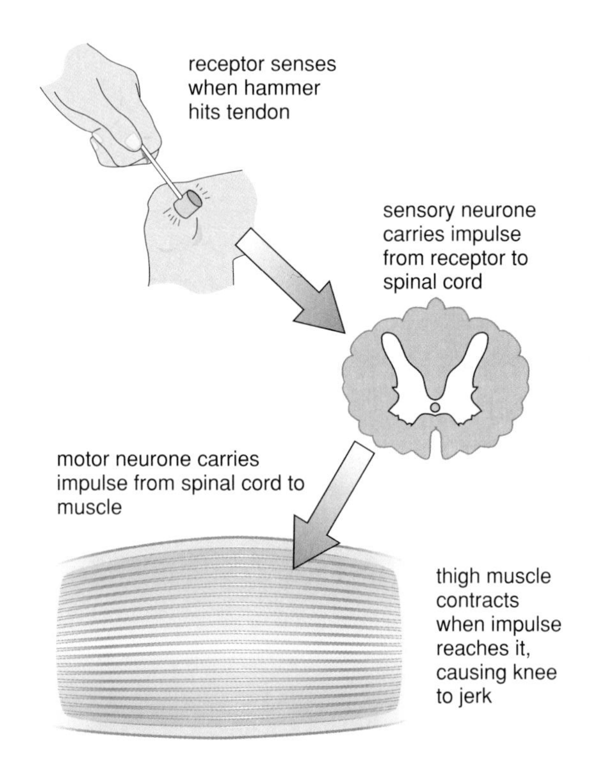

a Say whether each of the following actions is a reflex action or something we have learned to do:

- **i** Blinking when dust lands on the eye.
- **ii** Catching a tennis ball.
- **iii** Jumping when standing on a pin.
- **iv** Sneezing.
- **v** Putting your books away as soon the bell rings.

Pain-withdrawal reflexes

Many of the reflex actions in the body protect it from danger – they prevent the body from being injured.

Some people suffer from a rare condition where the temperature-change receptors in the skin don't work. If you picked up a very hot plate at lunchtime, you would probably drop it straight away. But a person with faulty temperature-change receptors would keep hold of it.

b Explain why you would drop the plate, but a person with faulty temperature-change receptors would not.

Glands in reflexes

Not all our reflexes depend on muscle movement. Other reflexes help the body to work properly. If you are queuing up for school lunch and you smell food that you like, your mouth will begin to water. In this case it is not a muscle that brings about the reaction, but the salivary **glands** in the mouth. This is the mouth-watering reflex.

Did you know?

Hiccups are caused by a reflex action. If you eat too much, pressure of your stomach on the diaphragm causes a reflex action.

Questions

1 Copy and complete the following sentences using the words provided:

automatic **receptor** **nerves**

A reflex action is an __________ reaction to a stimulus. The stimulus is detected by a __________.

Electrical impulses travel towards the brain or spinal cord via __________.

2 If you step on a pin, you automatically lift your foot away from the pin.

a What type of receptor is sensitive to the pin?

b What sort of reaction is this?

3 List as many reflex reactions as you can think of.

Summary

In a reflex action:

- Reflex actions are automatic responses to stimuli.
- A muscle responds by contracting, and a gland by secreting chemical substances.

Reflex actions help to protect the body from pain or injury.

2:3 The eye

Our eyes are sense organs that allow us to see the world. The different tissues of the eye work together to create an image of the world around us. This image is detected by cells and the information is passed to our brains along the optic nerve.

The diagram below shows the parts of the eye and their jobs.

Light scattered by objects around us enters our eyes through the **cornea**. It then passes through the pupil and is focused by the **lens**. The light needs to form a clear, focused image on the retina.

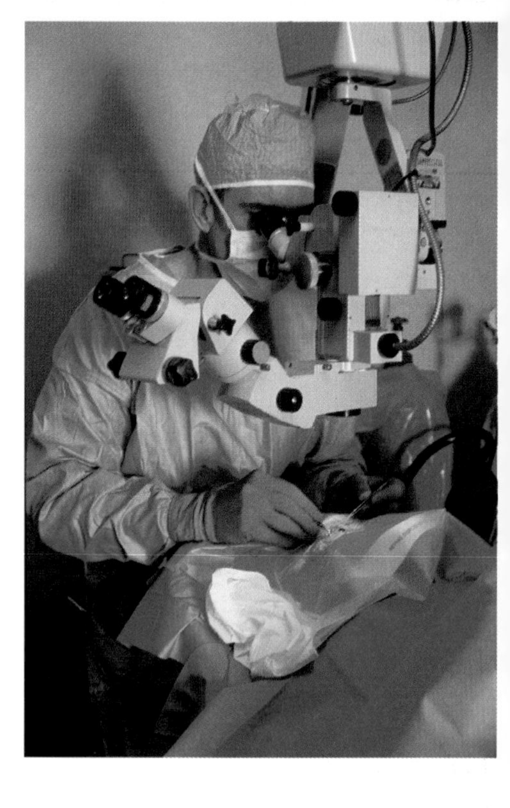

The photograph shows one of the most successful operations carried out today. Every year, in Britain alone, thousands of people partially lose their sight due to cataracts. As we get older the lenses in our eyes sometimes become cloudy and we cannot see clearly. The cloudy region of a lens is called a cataract.

Now, doctors can replace the natural lens by a plastic lens that works just as well. The patient in the photo is having his lens replaced.

a What is the job of a lens?

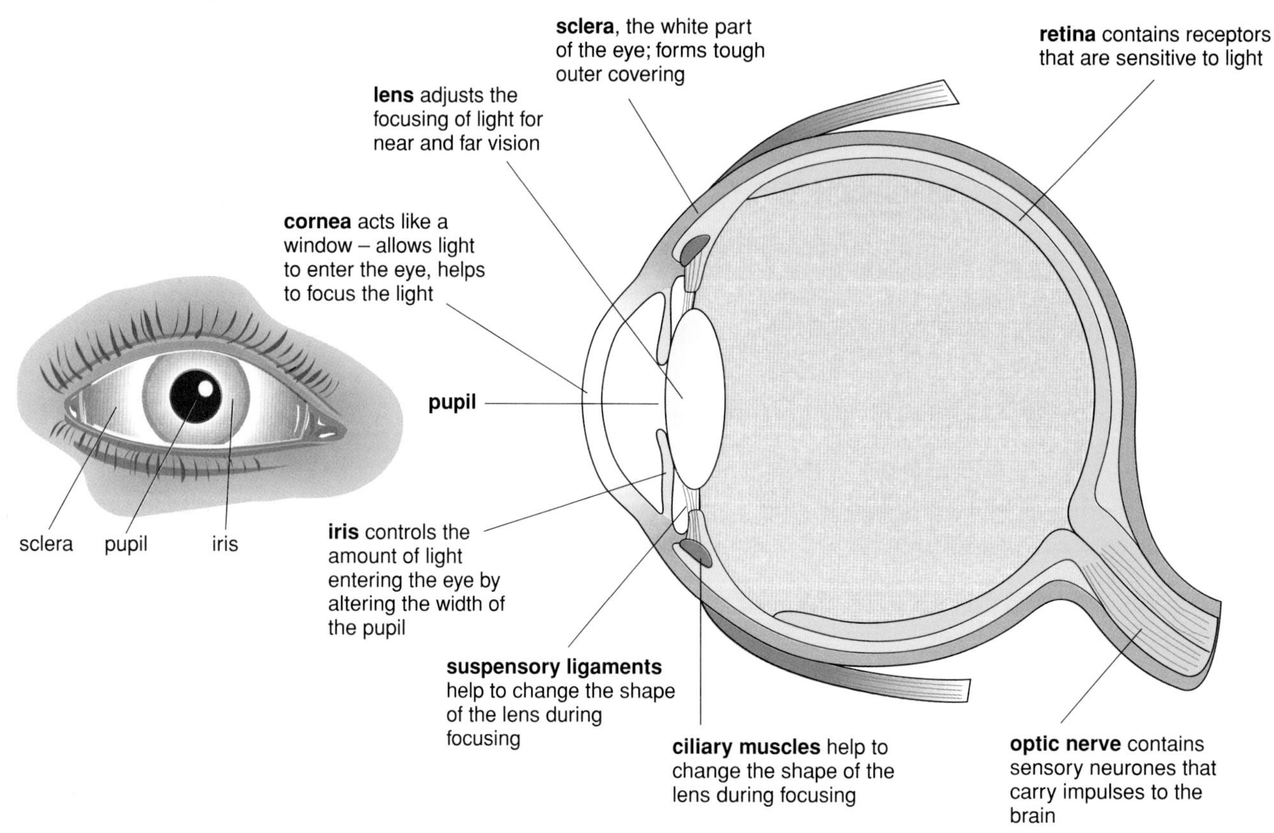

Focusing

The width of the lens can be changed by the **ciliary muscles** and the **suspensory ligaments**. To focus on near objects the lens is made thicker; to focus on distant objects the lens is made thinner.

ciliary muscle
focus on retina
suspensory ligaments
light rays from close-up object
fat lens

(a) near vision

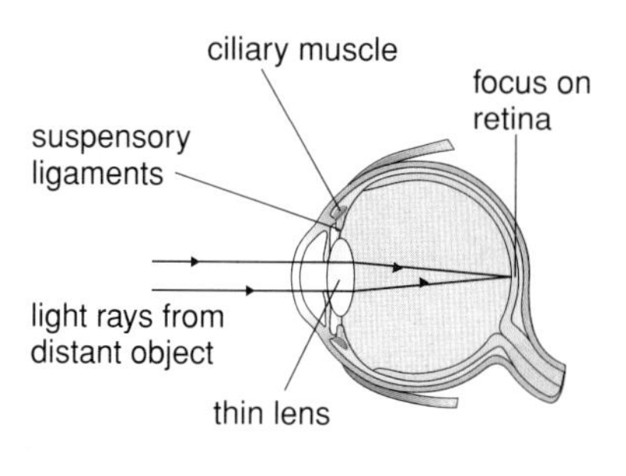

(b) far vision

Seeing in the dark and in the light

Look at your pupils in a mirror. Cover your left eye with your left hand. What happens to your right pupil in the next few seconds?

Now take the hand away from your left eye. What happens to both your pupils over the next few seconds?

b Why do you think this happened?

The pupils widen in the dark to let more light into the eyes. They become narrower in bright light to prevent too much light entering the eyes. The muscles in the iris change the width of the pupil.

in the dark

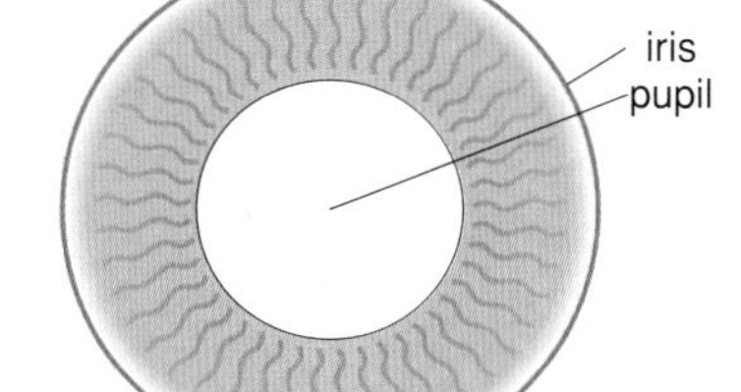

in the light

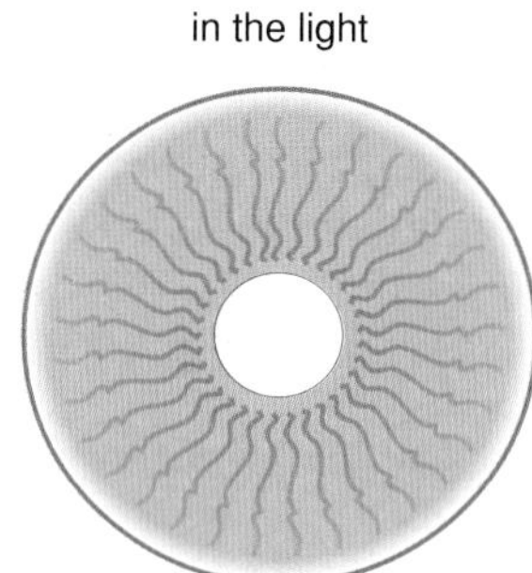

Sending information to and from the brain

The back of the eye contains the receptor cells that are sensitive to light. This layer of cells is called the **retina**. Very bright light can damage the retina; that is why you must never look directly at the sun.

The light-receptor cells of the retina are connected to the **optic nerve**. The optic nerve carries electrical impulses from the light-receptors to the brain. The brain then makes sense of these electrical impulses.

Questions

1 List in order the structures that light passes through on its way from the air to the retina.

2 List the jobs of the sclera, cornea, iris, ciliary muscles, lens, retina, optic nerve.

3 Copy and complete the following sentences using the words provided:

iris **pupil** **ciliary muscle**
suspensory ligaments **retina**

When we go into a dark room, the __________ gets smaller and the __________ gets bigger to let more light into the eye. When we look up from reading a book to see the TV, the shape of the lens is changed by the __________ __________ and the __________ __________. This is so that the picture from the TV screen is focused on the __________.

Summary

- Light enters the eye through the cornea.
- The cornea and the lens produce an image on the retina.
- The receptor cells in the retina send impulses to the brain along the optic nerve.
- The ciliary muscles and suspensory ligaments alter the shape of the lens to focus objects at different distances.
- The iris alters the size of the pupil to control the amount of light entering the eye.

2:4 Keeping a balance

Sports drinks have become big business in recent years. At any sports event you will see the players clutching their sponsored bottles.

Here is a recipe to make your own sports drink:

- 200 cm^3 of orange squash;
- 1 litre of water;
- a pinch (1 gram) of salt.

Athletes drink sports drinks to replace the fluid they lose as sweat when they exercise. If you have ever tasted sweat you will know that it is a bit salty. Sports drinks replace the water and salt that we lose in sweat. They also contain sugar; in the recipe above there is sugar in the orange squash.

a Why do athletes need sugar?

When you have to give more than you've got

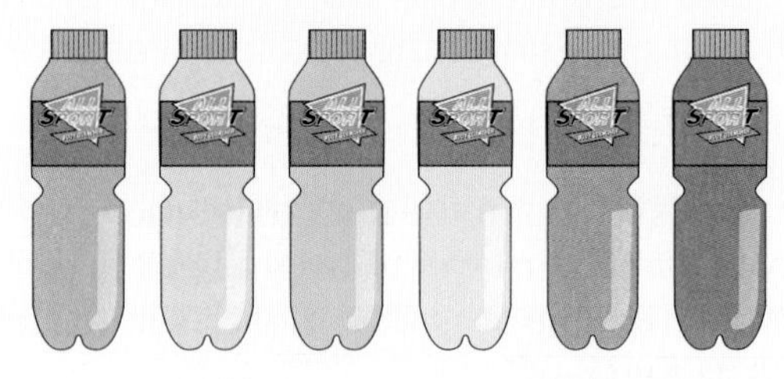

ALL SPORT **in the *NEW* "Claw" Bottle**

Peak performance

If athletes don't replace the water and salt they lose, their performance declines and they might even collapse. Why is this?

When we sweat we lose a lot of water, but not quite so much salt. This leaves us with more salt in our blood than is normal. If the balance of salt and water changes in our bodies, cells do not work so well. Sports drinks help to replace both the water and the salt.

Drinking a sports drink after exercise helps to keep the amount of salt in the athlete's blood constant. The body works at its best when conditions inside are kept constant. Keeping things inside the body constant is known as **homeostasis**.

b Can you think of any other conditions inside the body that need to be kept constant?

Diffusion

What happens to body cells if there is more salt than normal in the blood?

Molecules move from a region of high concentration to a region of lower concentration. This is called diffusion.

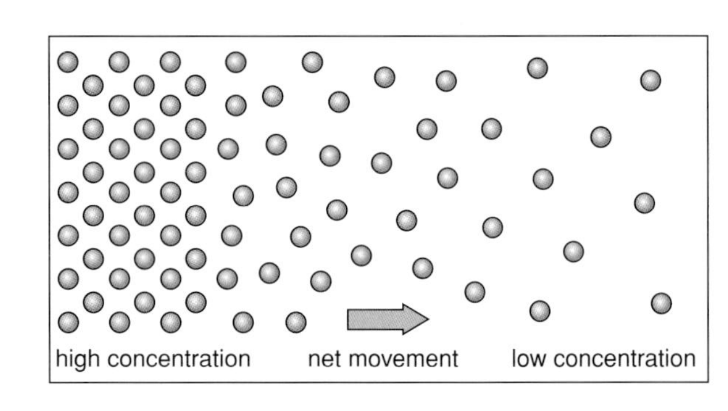

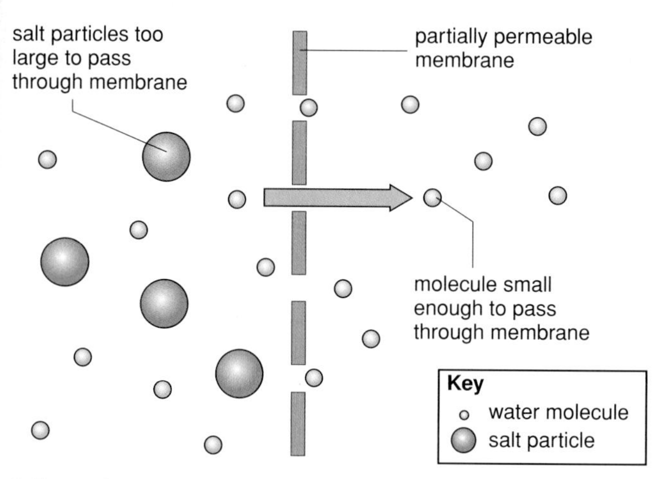

Cell membrane is partially permeable.

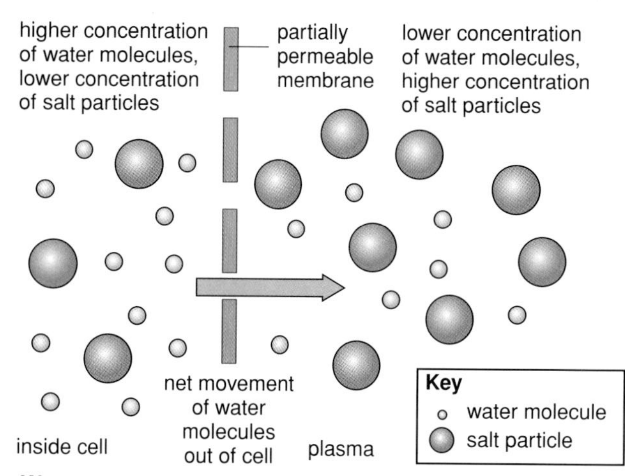

Water molecules move from high to low concentration through a partially permeable membrane.

Osmosis

Before sweating, the concentration of water molecules and salts is the same in both cells and blood. After sweating, there is a lower concentration of water in the blood than inside the cells. Water molecules diffuse out from the cells through the cell membrane. Diffusion of water molecules from the drink will help to keep the concentration of salts in the blood constant.

The cell membrane is **partially permeable**. Water molecules are small so they pass through the membrane, but the salt gets left behind. This process is called **osmosis**.

c What happens to the volume of the cells when water molecules diffuse out?

Questions

1 Copy and complete the following sentences using the words provided:

homeostasis diffusion partially permeable

Keeping conditions inside the body constant is called __________. Movement of a molecule from a high concentration of the molecule to a low concentration of the molecule is called __________. A membrane that lets small molecules through, but not large ones, is called __________ __________.

2 What will happen to the concentrations of water and salts in the blood after drinking a sports drink?

3 You have probably seen patients attached to 'drips' in hospital programmes on TV. The drip delivers a solution straight into the blood in a vein. The drip usually contains a drug dissolved in water. It also contains salt.

Explain why it is important that the drip contains salt.

Summary

- Homeostasis is keeping internal conditions in the body constant.
- Diffusion is the spreading of particles resulting in a net movement from a region of higher concentration of particles to a region of lower concentration.
- The greater the difference in concentration, the faster the rate of diffusion.
- Osmosis is the diffusion of water from a dilute to a more concentrated solution through a partially permeable membrane that allows the passage of water molecules but not the solute particles.

2:5 Getting rid of waste

Carbon dioxide

In cars and lorries, fuels are burned to provide the energy for movement. This uses fuel and oxygen, and releases waste products. The main waste products in a car's exhaust fumes are carbon dioxide and water.

Respiration in our body cells is similar to combustion. In our case the fuel is glucose. This reacts with oxygen in our cells to release energy. The waste products are carbon dioxide and water.

$$\text{glucose} + \text{oxygen} \rightarrow \text{carbon dioxide} + \text{water} + [\text{energy}]$$

We have to remove these waste products from our bodies. Both carbon dioxide and water leave the body in the air we breathe out. You may have noticed water condensing when you breathe onto a cold mirror.

a Water is also lost by other parts of the body. Can you think of two other ways in which the body loses water?

Urea

Our body also needs amino acids for building new cells. We get these amino acids by digesting the protein in our food. The body cannot store amino acids, so it has to get rid of excess amino acids – the ones it cannot use immediately.

Like glucose, amino acids contain energy. The **liver** splits the excess amino acids into two parts: an acid part and an amino part. The *acid* part is used in respiration, and the *amino* part is made into **urea**.

Urea is a poisonous waste product so the body has to get rid of it in the following way:

1. The urea is transported by the blood to the **kidneys**.
2. The kidneys extract urea, excess water and excess ions from the blood to make **urine**.
3. Urine flows from the kidney to the **bladder**.
4. Urine is stored in the bladder until we release it when we urinate.

Urine consists of:

- urea;
- excess water (water the body does not need);
- ions that the body does not need.

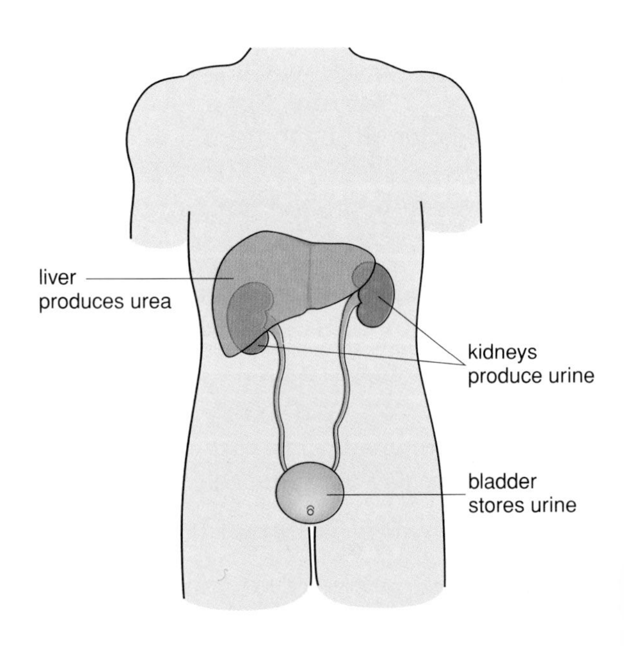

Kidneys

The kidneys regulate the amount of water in the body. To stay healthy the body needs to balance water gain and water loss. We call this the body's water budget.

The diagram shows how the water entering the body should balance the water leaving the body. When this happens we say that the water budget is in balance.

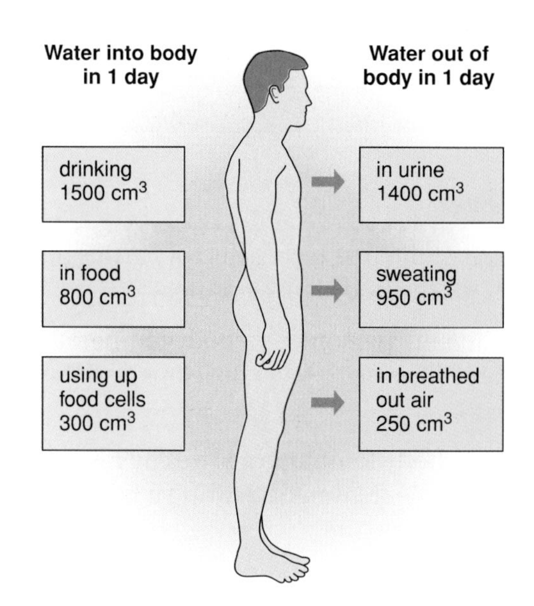

When kidneys fail

If the kidneys fail then waste materials build up in the body and begin to poison it.

There are two methods of treating people with kidney failure. The best method is to remove the failed kidneys and replace them with a healthy kidney. The healthy kidney can be transplanted from a person who has just died.

There is a big shortage of kidneys available for transplants. When no kidneys are available, patients are connected to a mechanical kidney called a **dialysis machine**.

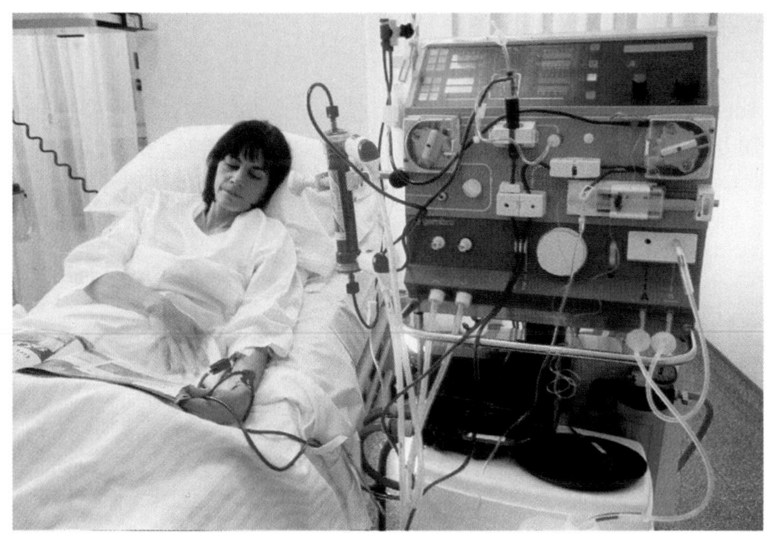

The patient is being treated with a dialysis machine.

Summary

Waste products which have to be removed from the body include:

- Carbon dioxide which is produced during respiration and leaves the body via the lungs.
- Urea which is produced in the liver by the breakdown of excess amino acids and is removed by the kidneys in the urine.
- Urine is temporarily stored in the bladder.
- Water leaves the body via the lungs when we breathe out, via the skin when we sweat, and excess is lost via the kidneys in the urine.
- Ions are lost via the skin in sweat, and excess ions are lost via the kidneys in the urine.

Questions

1 Copy and complete the following sentences using the words provided:

respiration	**lungs**	**amino acids**	**liver**
kidneys	**bladder**	**water**	**ions**

Carbon dioxide is a waste product of __________. It leaves the body via the __________. Urea is formed by the breakdown of __________ __________ in the __________. Urine is produced by the __________ and stored in the __________. Urine consists of urea, excess __________ and excess __________.

2 Describe two important jobs that the kidneys perform.

3 On a hot day your body will lose more water than it would on a cold day. Describe two ways in which you could balance the water budget.

4 Explain why a cold mirror 'steams up' when you breathe on it.

2:6 Controlling body temperature

Fever

One of the first tests a doctor carries out on a patient is to check the body temperature.

A healthy person has a body temperature of about 37°C. In some illnesses a fever develops and the body temperature rises.

If the body temperature rises by more than 4°C, then doctors must give treatment to bring it back down before permanent damage is done.

What happens when you get too hot?

Have you ever thought why egg white turns from liquid to solid when you boil the egg? No matter how hard you try, you can't turn solid egg white back into a liquid. This is because egg white is a protein.

High temperatures **denature** proteins. The egg white has become denatured when it is boiled. Many parts of the body cells are made of protein, so if the cell's proteins are denatured, the cell won't work as well.

Constant body temperature

Mammals and birds are the only groups of animals that can keep their body temperature constant. Birds have a rather higher temperature than mammals, about 40°C.

a Why do you think we need a constant body temperature?

Swallows migrate to Africa for our winter. Swallows feed on flying insects, and there are very few flying insects in Britain in winter.

So where do flies go in winter? An insect's temperature is the same as that of the environment. In winter insects are too cold to move about so most of them die or hibernate. That's why swallows have to go to warmer climates to find food.

Animals with a constant body temperature can remain active all the year round.

Raising the temperature

Our bodies have a lot of chemical reactions going on. Enzymes are proteins that make reactions go more quickly. Enzymes control most chemical reactions in cells.

Enzymes work best at a particular temperature. When we are healthy our bodies are at 37°C. This is the temperature at which our enzymes work best. If our bodies were cooler then 37°C the chemical reactions in our cells would be much slower. If we heat enzymes above 45°C they get broken up and stop working. They become denatured. Denatured enzymes are removed from our bodies as urea.

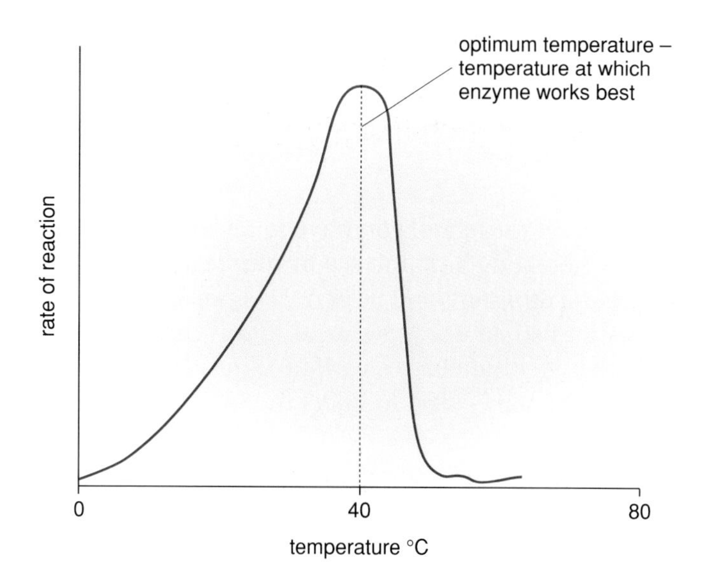

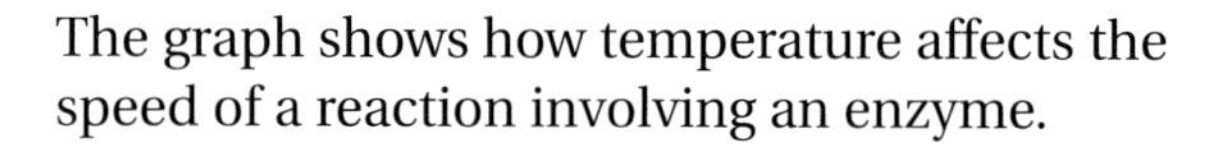

The graph shows how temperature affects the speed of a reaction involving an enzyme.

b Why does the rate of enzyme controlled reaction:

i Rise between 0 and 40°C?

ii Fall rapidly above 45°C?

Keeping warm

In Britain, the outside temperature rarely rises above the mid twenties Celsius. Our body temperature is at least 10°C higher than this. Where does the heat come from?

All the cells in our body respire. Some of the energy from respiration is released as heat. Much of the heat for our bodies comes from the liver. The liver is the body's chemical factory – hundreds of chemical reactions take place there 24 hours per day. A lot of the heat for our bodies comes from the liver.

c Which parts of the body produce most heat during exercise?

	Water loss (cm^3)	
	cool day	hot day
breath	400	
skin	500	
urine	1500	
faeces	150	

Four ways water leaves the body on a cool day.

Sweating

We get very hot during exercise, producing a lot of sweat. Our bodies sweat to keep us cool. Sweat is made by tiny glands in the skin. Sweating by itself does not cool us down, it is the evaporation of sweat that cools us down. Heat energy needs to be supplied to evaporate a liquid. Evaporating sweat uses heat energy from our bodies, leaving us cooler.

Questions

1 The table above shows four ways in which water leaves the body on a cool day.

Copy the table, then fill in the right hand column to show whether the amount of water lost on a hot day would be less, more, or the same. Give the reason for each of your answers.

2 Describe the different ways in which our bodies maintain the correct temperature.

3 Explain why it is important that our bodies stay at 37°C.

Summary

- Mammals and birds maintain a high, constant body temperature.
- Enzymes work best at this temperature.
- Heat is released when body cells respire.
- Sweating cools the body when sweat evaporates.
- Water and ions are lost when we sweat.

2:7 Controlling blood sugar

When you set off for school you have to remember your books and a pen. There may be someone in your year who also needs to remember a different kind of pen. This pen contains a drug called **insulin**. People who have a condition called **diabetes** may have to inject themselves with insulin several times per day. The insulin pen is used to inject the insulin.

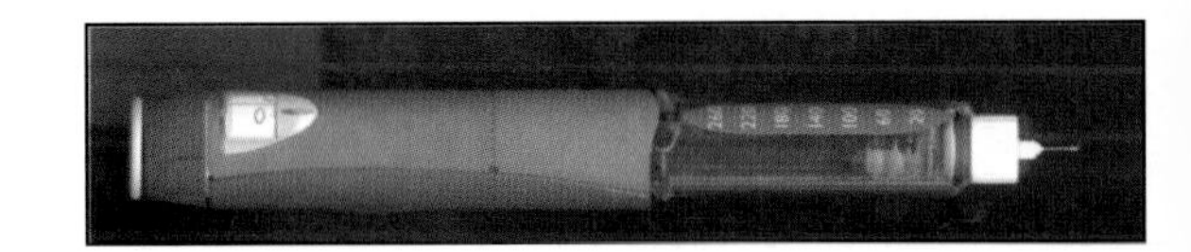

Insulin

Healthy people make their own insulin. It is produced by the pancreas.

Our bodies digest carbohydrates into glucose. This glucose is absorbed into the bloodstream. So, after a meal the concentration of glucose in the blood rises. If the concentration of glucose in the blood gets too high it can do you harm.

Cells in the pancreas detect rises in blood glucose concentration. They respond by producing insulin.

Insulin affects all the cells in the body. It makes them take in glucose so the concentration of glucose in the blood comes down. The amount of insulin produced is controlled by cells in the pancreas, not by the brain.

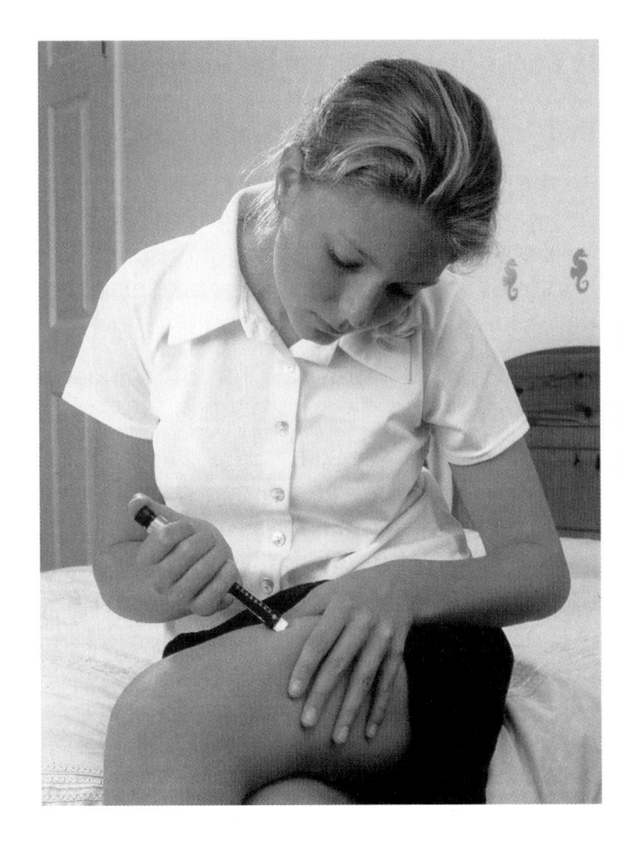

Hormones

Insulin is one of a group of chemicals called **hormones**. Hormones control many of the processes in the body.

They are produced in organs called glands. The pancreas is a gland. Hormones are secreted by glands into the blood and then transported around the body. Each hormone only affects particular organs. The organs that are affected by hormones are called **target organs**.

a **The pancreas also secretes enzymes for digestion. Where do these enzymes go as they leave the pancreas?**

Diabetes

In people with diabetes, cells in the pancreas do not secrete enough insulin.

The majority of people who have diabetes can control the condition by watching their diet. They have to be careful about the amount of sugar-containing foods that they eat.

b **Explain why people with diabetes must not eat very much sugar-containing food.**

Some people suffering from diabetes have so little insulin that they have to make up for the loss of insulin by injecting themselves with it several times a day.

c Suggest why people with diabetes inject small amounts of insulin several times a day rather than just injecting one large dose of insulin.

Did you know?

Insulin was not discovered until 1921. Its production by the drug industry saved the lives of millions of people who would certainly have died from the severe form of diabetes. Until recently, insulin for injections was extracted from the pancreas of pigs, sheep and cattle that were killed for meat. Now, some human insulin is made by genetically engineered bacteria.

Getting the balance right

The pancreas secretes another hormone called **glucagon**. Glucagon has the opposite effect to insulin. If the blood glucose concentration drops too low, the pancreas secretes glucagon. Glucagon causes the blood glucose concentration to rise to normal levels.

d Give two reasons why the blood glucose concentration might fall.

Questions

1 Copy and complete the following sentences using the words provided:

hormones pancreas insulin glucagon diabetes

Chemicals called __________ control many of the processes in the body. The concentration of sugar in the blood is monitored by the __________. If there is too much sugar in the blood the pancreas secretes __________. If there is too little sugar in the blood the pancreas secretes __________. If the pancreas does not secrete enough insulin, a person suffers from __________.

2 People with diabetes need to avoid eating snacks with lots of sugar in them. Why do you think this is?

3 Two people, A and B, were given an identical meal containing large amounts of glucose. The concentrations of glucose and insulin in their blood plasma were measured at regular intervals over the next 150 minutes. The graphs show the results.

a Which person, A or B, had diabetes? Give the reason for your answer.

b Explain why the blood glucose concentration of person B rose for one hour after the meal, then fell.

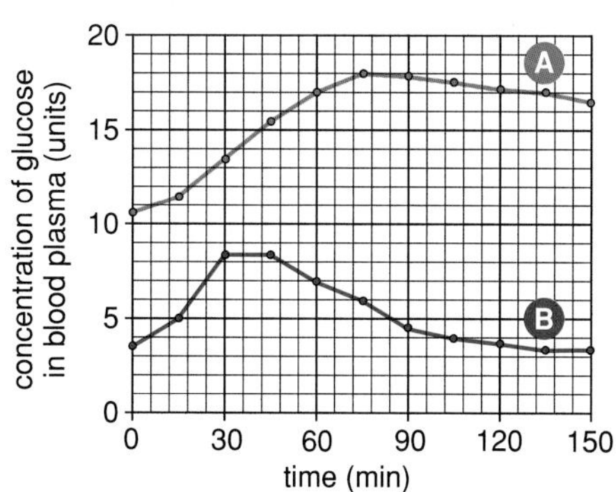

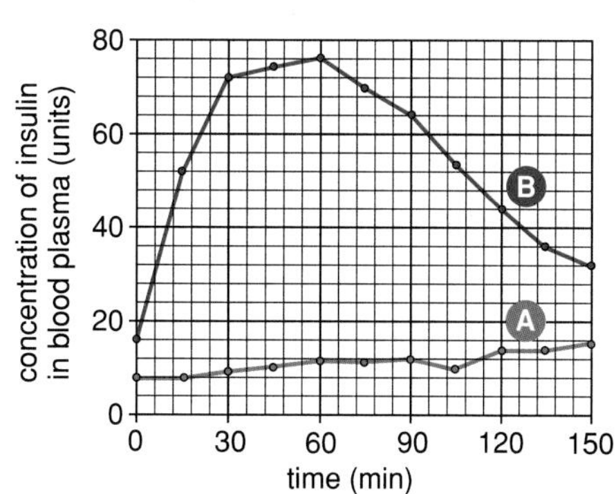

Summary

- Hormones are secreted by glands and are transported to their target organs by the bloodstream.
- Hormones control many body processes.
- The blood glucose concentration is controlled by the hormones insulin and glucagon which are secreted by the pancreas.
- Diabetes is a disease in which blood glucose concentration may rise because the pancreas does not secrete enough insulin.
- Diabetes may be treated by careful attention to diet and by injecting insulin into the blood.

2:8 Drugs

Drugs are often in the news or being talked about by people. It seems that all these people have different attitudes towards drugs. There are thousands of different drugs in the world; some have been useful whilst others have caused nothing but problems.

a What do you think the people in the above picture mean when they talk about 'drugs'?

What is a drug?

Drugs are chemicals that affect the way your body works. Some drugs may help you when you are ill while others may change the way you think or feel pain.

Some drugs are illegal and others can be bought in your local shop.

b Make a table with three columns. From the list below, group all the products containing drugs that are illegal, those that can only be bought by adults or with a prescription, and those that anyone can buy.

aspirin, heroin, alcohol, coffee, cola, cannabis, antibiotics, cigarettes, tea, cough sweets

All drugs change the chemical reactions in our bodies. When your body gets used to the change it may become **dependent** on the drug. Your whole body chemistry may have changed. When this happens you have become **addicted** to drugs.

All of the products in question b contain drugs; some of them are very addictive whilst others are less addictive.

Drugs like heroin and nicotine (the drug in cigarettes) are very addictive. People may become addicted to these drugs very quickly.

When people who are addicted stop taking drugs their bodies may lose their natural balance. This can make them very ill or unhappy for a while and is called a **withdrawal symptom**.

Why do people take drugs?

People take drugs for lots of different reasons. If your body is not carrying out its normal chemical processes a doctor may advise you to take a particular drug to help. If you were feeling a little tired you might drink a cup of coffee to wake you up.

All drugs may damage your health in some way, but a doctor can help you decide if you need to take a drug.

Sometimes people just use drugs to change their mood, for instance, drinking alcohol or smoking cigarettes. People who smoke and drink need to understand the risks they are taking.

Other drugs are so dangerous that they are illegal in Britain. Heroin is highly addictive and causes lots of health problems. Cannabis is another illegal drug that causes lung cancer.

It is not only illegal drugs that are dangerous. Cigarettes and alcohol cause thousands of deaths in Britain every year. Many people would argue that these drugs should be made illegal.

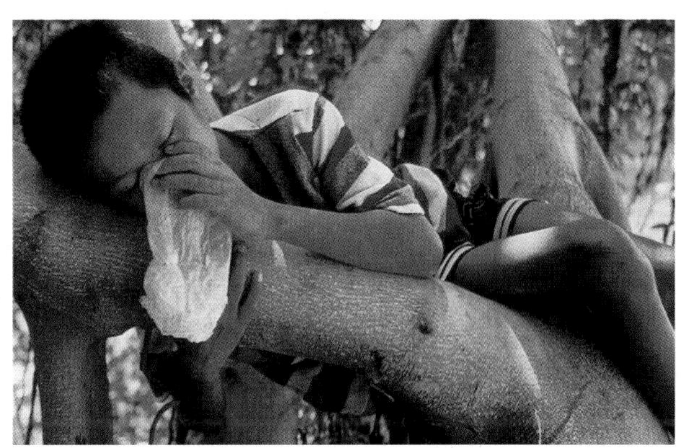

Sniffing glue can be fatal.

Solvents

Some people sniff **solvents** to get a 'high'. These solvents include glue, lighter fuel, petrol and aerosols.

For some people the effects of solvents are similar to alcohol. Others get visions and hallucinations.

Sniffing solvents is both dangerous and addictive. Solvents damage the brain and the liver. In Britain about 125 young people die each year due to damage caused by solvents.

Questions

1 Copy and complete the following sentences using the words provided:

chemical reactions **dependent**

withdrawal symptons **brain** **liver**

Drugs change some __________ __________ in our bodies. As a result of these changes we become __________ on the drug. If we stop taking the drug we will experience __________ __________.

Solvents cause damage to our __________ and __________.

2 Explain why some people choose to take illegal drugs, or drugs that may harm them.

3 Design a leaflet to help people your age make the best decisions about drugs. You should explain how drugs can harm or kill you.

Summary

Drugs:
- Change the chemical processes in people's bodies.
- May cause them to become dependent or addicted.
- May cause them to suffer withdrawal symptoms without them.

Solvents:
- Affect behaviour.
- Damage the lungs, liver and brain.

2:9 Alcohol and tobacco

Cigarettes

Cigarettes are made from dried tobacco leaves. Tobacco smoke is a mixture of hundreds of different chemicals. Many of these chemicals are dangerous to the human body. Some chemicals stimulate cells to divide rapidly. This leads to cancer. Tobacco affects the lungs in other ways as well.

Bronchitis

Some chemicals paralyse the tiny hairs, called cilia, that line the air passages, and cause the mucus glands to produce more mucus than they should. After a while mucus builds up in the lungs making it hard to breathe and causing a 'smokers' cough'. Coughing too much can cause further damage to the lungs.

Sometimes a smoker's **bronchioles** (small tubes in the lungs) will become narrowed. This makes breathing harder and the smoker may become out of breath.

Emphysema

The chemicals in tobacco can also cause parts of the lungs to become less flexible. Emphysema is a condition where the alveoli lose their flexibility. This will make breathing difficult.

Carbon monoxide

One of the poisonous substances in tobacco smoke is **carbon monoxide**. This prevents red blood cells from carrying oxygen – so a smoker's blood carries less oxygen. This is one reason why babies born to women who smoke are lighter at birth on average than babies born to non-smokers.

The heart and blood vessels

Some of the chemicals in tobacco smoke cause disease of the heart and blood vessels. Many smokers die of **heart attacks** caused by the blocking of the arteries that supply blood to the heart muscles.

Did you know?

People have been smoking tobacco in Britain for nearly 500 years. In the 1950s doctors decided that cigarettes were causing problems for smokers. Many of the doctors gave up smoking at once. For the next twenty years people studied the health of doctors and found that the ones who stopped smoking became much less likely to die of lung cancer. Also in the 1950s enough evidence had been gathered to prove that chemicals in tobacco caused lung cancer. Smokers are about 30 times more likely to get lung cancer than non-smokers.

Alcohol

Alcohol is the oldest known drug. It is a drug that affects the nervous system. Small amounts cause excitement and lack of inhibitions. But even small amounts slow down the body's reactions. That is why it is illegal to drive a car with only a small amount of alcohol in the blood.

Larger amounts of alcohol lead to lack of self-control. One example of this is the fights that often break out when young adults, 'lager louts', drink too much.

Drinking far too much alcohol can cause a person to lose consciousness and even go into a coma. Every year we hear a tragic story of someone dying from drinking too much alcohol on a stag night.

Drinking too much alcohol will lead to alcohol addiction. Many alcohol addicts die early because of damage to their liver or brain. It is important that people who drink understand when they are likely to do damage to themselves.

The National Health Service recommends that a fully grown male drinks no more than 4 units (about 2 pints of beer) of alcohol per day; the figure for an adult female is 2 units. Alcohol will have a more dangerous effect on people under 18, so it is illegal for them to buy alcohol in Britain.

The amount of alcohol in the bloodstream can be worked out from this simple test.

Questions

1 **Copy and complete the following sentences using the words provided:**

nicotine	**quickly**	**lung**	**bronchitis**
mucus	**alveoli**	**liver**	**brain**
reactions	**nervous**	**carbon monoxide**	

The addictive substance in tobacco is called __________. A cancer results when cells divide too __________. Cigarette smoke causes cancer of the __________. The disease which results from paralysis of the tiny hairs in the lungs is called __________. The cells of the lungs of a person with this disease produce too much __________. Emphysema is a condition where __________ lose their elasticity. The gas in cigarette smoke that affects red blood cells is __________ __________.

Alcohol affects the __________ system by slowing down our __________. Drinking too much alcohol can damage our __________ and __________.

2 **Chemicals in tobacco smoke also cause the bronchioles to narrow. The symptom for this is that the person becomes 'out-of-breath'. Suggest why a person with narrower bronchioles becomes 'out-of-breath'.**

Summary

Alcohol:

- Affects the nervous system by slowing down reactions.
- Leads to lack of self-control, unconsciousness or even coma.
- Damages the liver and brain.

Tobacco smoke contains substances which can help to cause:

- Lung cancer.
- Other lung diseases such as bronchitis and emphysema.
- Disease of the heart and blood vessels.

Carbon monoxide reduces the oxygen-carrying capacity of the blood; this can deprive a foetus of oxygen and lead to a lower birth weight.

2:10 Plant cells

Plans are already underway to construct manned spacecraft to visit other planets in our solar system. The voyages in these craft will take years. A major problem is how to provide enough food and oxygen for such a long voyage.

The answer lies not in finding extra storage space but in using tiny plants to recycle human waste products into both food and oxygen.

a What waste products do humans breathe out?

The tiny plants that will be used are aquatic plants. Each of these plants consists of only one cell. They will be grown in tanks on the spacecraft.

Animal cells cannot produce food and oxygen, so what have plant cells got that animal cells have not?

Structure of a plant cell

These are plant cells as seen through a microscope. Apart from having a regular shape and a green colour you cannot see much further detail.

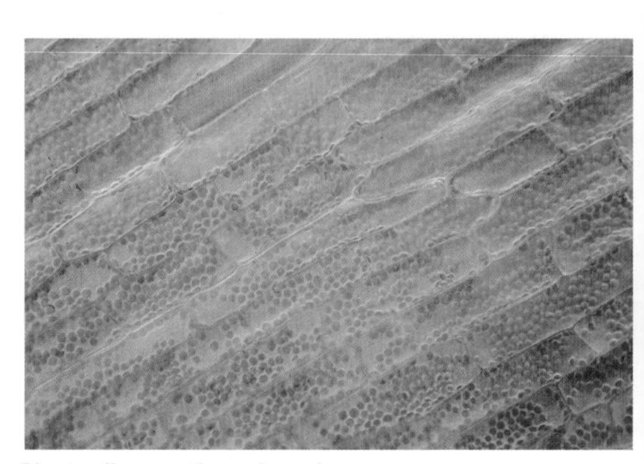

Plant cells seen through a microscope.

b List the differences between the plant cell and the animal cell.

It is one of these differences that enables plant cells to produce food and oxygen. Plant cells have chloroplasts. Chloroplasts are able to use energy from light to make food and oxygen.

In plant cells the nucleus, cytoplasm and cell membrane have the same job as in animal cells. The jobs of the parts of a plant cell are shown in the table.

Part of plant cell	Job
nucleus	controls the activities of the cell
cytoplasm	where many of the chemical reactions occur
cell membrane	controls the passage of substances in and out of the cell
cell wall	strengthens the cell
chloroplasts	absorb light to make food
permanent vacuole filled with cell sap	when full, helps to support the cell

c Explain how the cell membrane controls the passage of substances in and out of the cell.

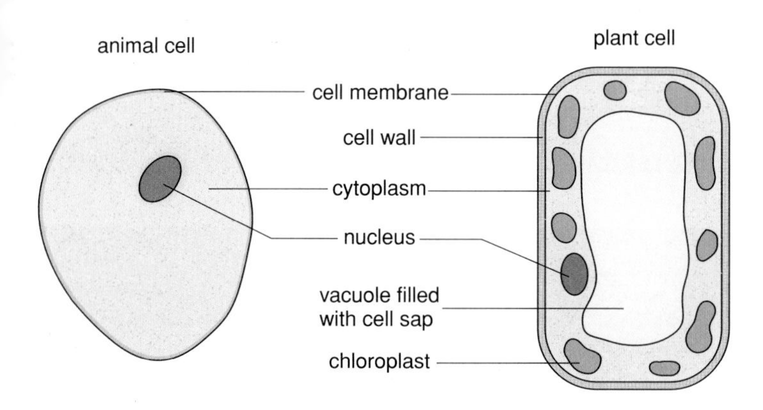

Chloroplasts are green because they contain the green pigment **chlorophyll**. It is this pigment that absorbs energy from light. Cells will only produce chlorophyll in the light. Plants grown in the dark have a yellow colour.

d Would you expect root cells to have chlorophyll? Give the reason for your answer.

Did you know?

The photograph shows the chemical factory of the future. This particular crop is oilseed rape. As the name suggests the plant produces oil. Scientists are now genetically engineering crop plants to produce many chemicals that are usually produced in factories.

Questions

1 Copy and complete the following sentences using the words provided:

membrane	**nucleus**	**chemical reactions**
food	**cell wall**	**supports**

The passage of substances in and out of a cell is controlled by its __________. The __________ controls the activities of a cell. The cytoplasm is where many __________ __________ take place. Chloroplasts make __________. A plant cell is strengthened by its __________ __________. The vacuole __________ the cell.

2 Can you think of the advantages of producing chemicals by growing crops rather than using factories? Are there any disadvantages?

Summary

- Parts of plant cells and their functions include:
  - the nucleus, controls the activities of the cell;
  - the cytoplasm, where most of the chemical reactions take place;
  - the partially permeable membrane, controls the passage of substances in and out of the cell;
  - the cell wall, strengthens plant cells;
  - chloroplasts, absorb energy from light to make food;
  - the large vacuole, filled with cell sap which helps to support the cell.
- Plants obtain carbon dioxide from the atmosphere

2:11 Photosynthesis

The photograph shows the traditional way of harvesting rice in Thailand. Rice is the staple diet of over half of the world's population. Rice is the fruit of grass plants. The fruit is packed with food needed for young grass plants to grow. The food is made by **photosynthesis** – the most important chemical process on Earth.

The photosynthesis equation

Photosynthesis consists of many chemical reactions, but these can be summarised by the word equation:

carbon dioxide + water + [light energy]
→ glucose + oxygen

Investigating photosynthesis

Plants use photosynthesis to make carbohydrates. If we want to find whether a plant has photosynthesised we need to test it for carbohydrates. The carbohydrates made by plants include glucose and starch.

It is not very easy to test leaves for glucose, so we test the leaves for starch. We use iodine solution to test for starch. If the leaf turns blue/black when tested, the leaf has made starch.

Is light needed for photosynthesis?

We can check this by doing an experiment. To do this we allow some parts of the leaf to receive light but not others. A foil stencil should be attached to the leaf as shown in the diagram. The plant is left in the light for a few hours, then the leaf is tested for starch.

The diagram shows the set-up and the results of this experiment. Only the parts of the leaves which were not covered turned blue in the iodine test.

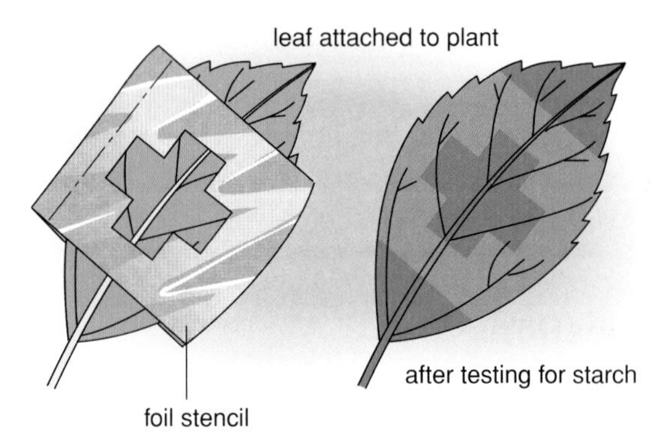

a Explain the appearance of the leaf after the starch test.

Is carbon dioxide needed for photosynthesis?

To test for this we must stop one leaf of a plant getting carbon dioxide. The diagram shows how this is done.

After a few hours in the light, the leaf may be tested for starch. The leaf in the flask did not turn blue in the starch test, but a leaf which had not been in the flask turned blue.

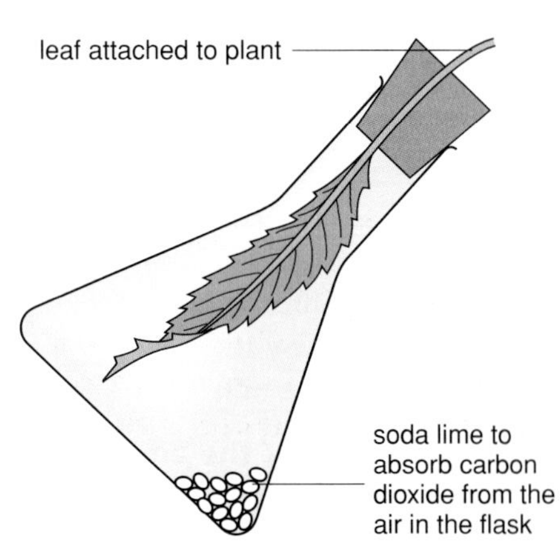

b What is the control in this experiment? Explain the results of the starch test.

Showing that chlorophyll is needed for photosynthesis

We can't take chlorophyll out of a leaf without killing it. But some leaves have patches without chlorophyll. These leaves are called **variegated** leaves.

c What colour would you expect the parts of the leaf without chlorophyll to be?

A plant with variegated leaves is left in the light for a few hours. One of its leaves is then tested for starch. The results of the starch test are shown in the diagram.

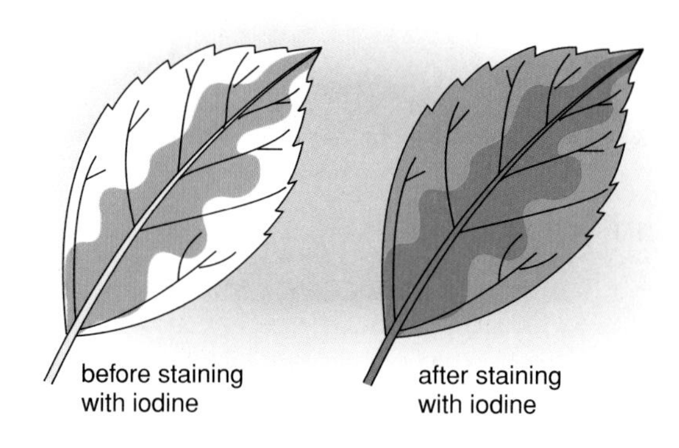

d Explain the results of the starch test.

These three experiments confirm that green plants need carbon dioxide, light and chlorophyll to make starch.

Oxygen

What gases do plants give out? We can collect the gas given out by plants using an aquatic plant in the equipment shown. Aquatic plants are ones that grow naturally underwater. They can obtain dissolved oxygen and dissolved carbon dioxide from the water. When the gas in the tube is tested we find that it is mainly oxygen when the plant is in the light.

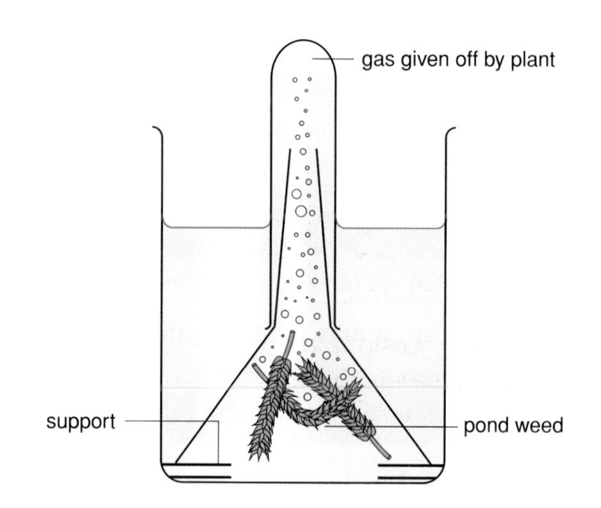

e How would you show that the gas in the tube was oxygen?

Questions

1 Copy and complete the following sentences using the words provided:

carbon dioxide	**water**	**chlorophyll**
glucose	**starch**	**oxygen**

To photosynthesise plants need to take in __________ __________ and __________.

In photosynthesis, light energy is absorbed by __________. The first carbohydrate produced in photosynthesis is __________. Some of this carbohydrate is converted into __________.

The gas given off in photosynthesis is __________.

2 Over two hundred years ago, a scientist called Joseph Priestley put a potted plant, a mouse and some mouse food in a large, airtight jar. He left them for several weeks. Both the plant and the mouse survived. Explain why.

Summary

During photosynthesis:

- Light energy is absorbed by chlorophyll.
- This energy is used to convert carbon dioxide and water into a sugar (glucose).
- Oxygen is released as a by-product.
- Some of the glucose is converted into starch.

2:12 Growing crops

Limiting factors

The photograph shows a crop of flowers in a greenhouse. Plants need special conditions to grow well. The conditions outside in fields cannot be easily controlled but in greenhouses it is easier to control conditions for the plants.

a Greenhouses are expensive to build. Can you think of reasons why farmers grow some crops in greenhouses rather than outside in fields?

The photosynthesis equation tells us what conditions plants need to grow.

carbon dioxide + water + [light energy] →
glucose + oxygen

Increasing the concentration of the chemicals can usually increase the speed of a reaction. Many reactions also go faster if temperature is increased. This will make the plants grow more quickly.

Changes that can increase the speed of a reaction are called **limiting factors**. A plant's growth will be slowed (limited) if any of these limiting factors are in short supply.

The factors that may limit the rate of photosynthesis are:

- low temperature;
- shortage of carbon dioxide;
- shortage of light.

Carbon dioxide as a limiting factor

The atmosphere has a very low concentration of carbon dioxide. This limits the rate of photosynthesis. We cannot improve the carbon dioxide supply to crops growing in fields but we can give plants growing in greenhouses more carbon dioxide.

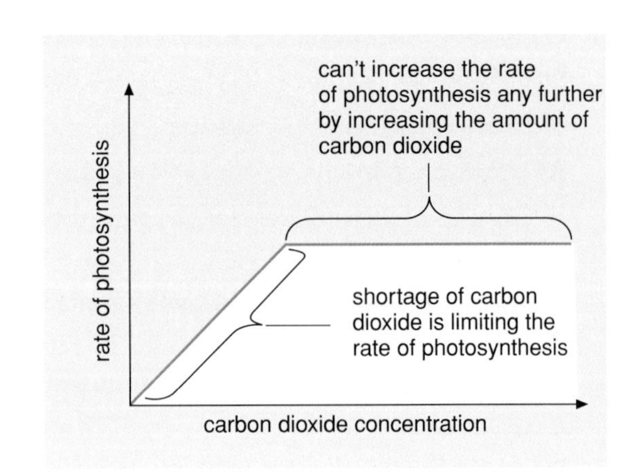

Growers usually increase the amount of carbon dioxide in a greenhouse by burning fossil fuels such as propane. They find that they can only increase the rate of photosynthesis up to a set level.

b Why is it easier to control the amount of carbon dioxide in a greenhouse when this level is reached?

Light as a limiting factor

Photosynthesis requires light, so plants do not photosynthesise in the dark, and they only photosynthesise slowly in dim light. We can give plants growing in greenhouses more light. This is not needed in summer, but it is often done in winter for valuable crops.

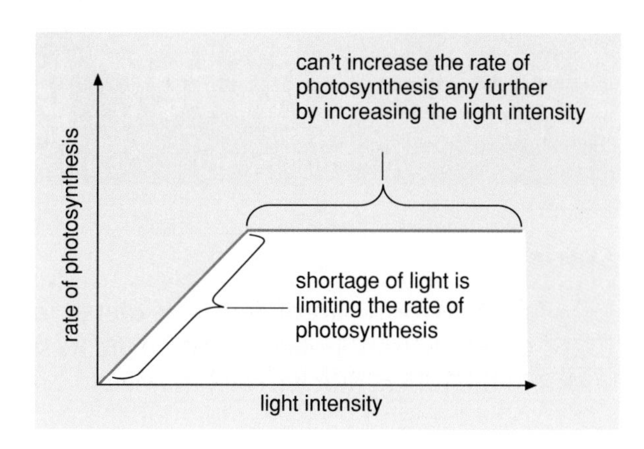

Temperature as a limiting factor

Temperature speeds up the rate of photosynthesis in two ways:

- it speeds up the movement of the molecules which react inside the chloroplasts;
- it speeds up carbon dioxide from the atmosphere getting into the leaves.

We cannot improve the temperature for crops growing in fields but we can for plants growing in greenhouses. This is not needed in summer, but it is often done in winter for valuable crops. The heat is usually supplied by burning propane.

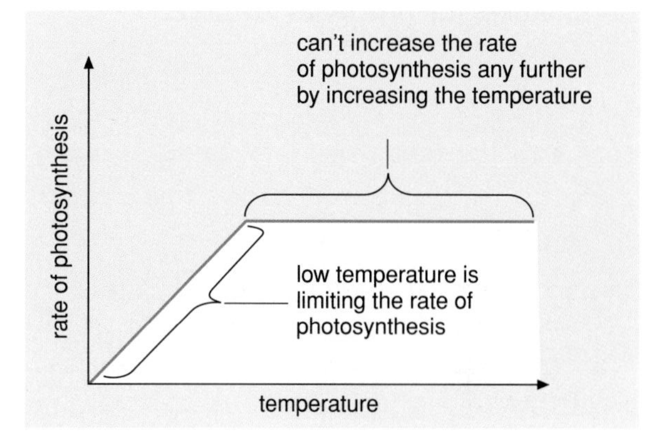

c Apart from cost, what is the advantage of heating a greenhouse by burning propane rather than using electric heating?

Questions

1 List three factors that may limit the rate of photosynthesis.

2 Copy and complete the following sentences using the words provided:

propane carbon dioxide low temperature

Growers can increase the amount of carbon dioxide in a greenhouse by burning ________.

At noon on a hot, sunny, summer day, the rate of photosynthesis is most likely to be limited by low ________ ________.

At noon on a sunny winter day the rate of photosynthesis is most likely to be limited by ________ ________.

3 Sketch a graph to show how the rate of photosynthesis of a crop plant will change through 24 hours of a day. Start the horizontal axis at midnight on one day and finish at midnight on the next day.

4 Gardeners at stately homes in Victorian times were asked to provide fresh fruit and vegetables in the winter. They did this by having beds of manure in small greenhouses. The manure gave off both heat and carbon dioxide. Can you think why?

Summary

The rate of photosynthesis may be limited by:

- Low temperature – this would happen in winter.
- Shortage of carbon dioxide – this might happen around noon on a sunny day.
- Shortage of light – obviously at night, but perhaps at dawn and dusk.

2:13 After photosynthesis

Storing carbohydrate

Chips seem to form part of the staple diet of most teenagers. Chips are made from potatoes. Potato plants store lots of starch in the tubers that we call potatoes.

Potato plants store starch to survive over winter, and to produce more potato plants next spring. The diagram below summarises the processes involved.

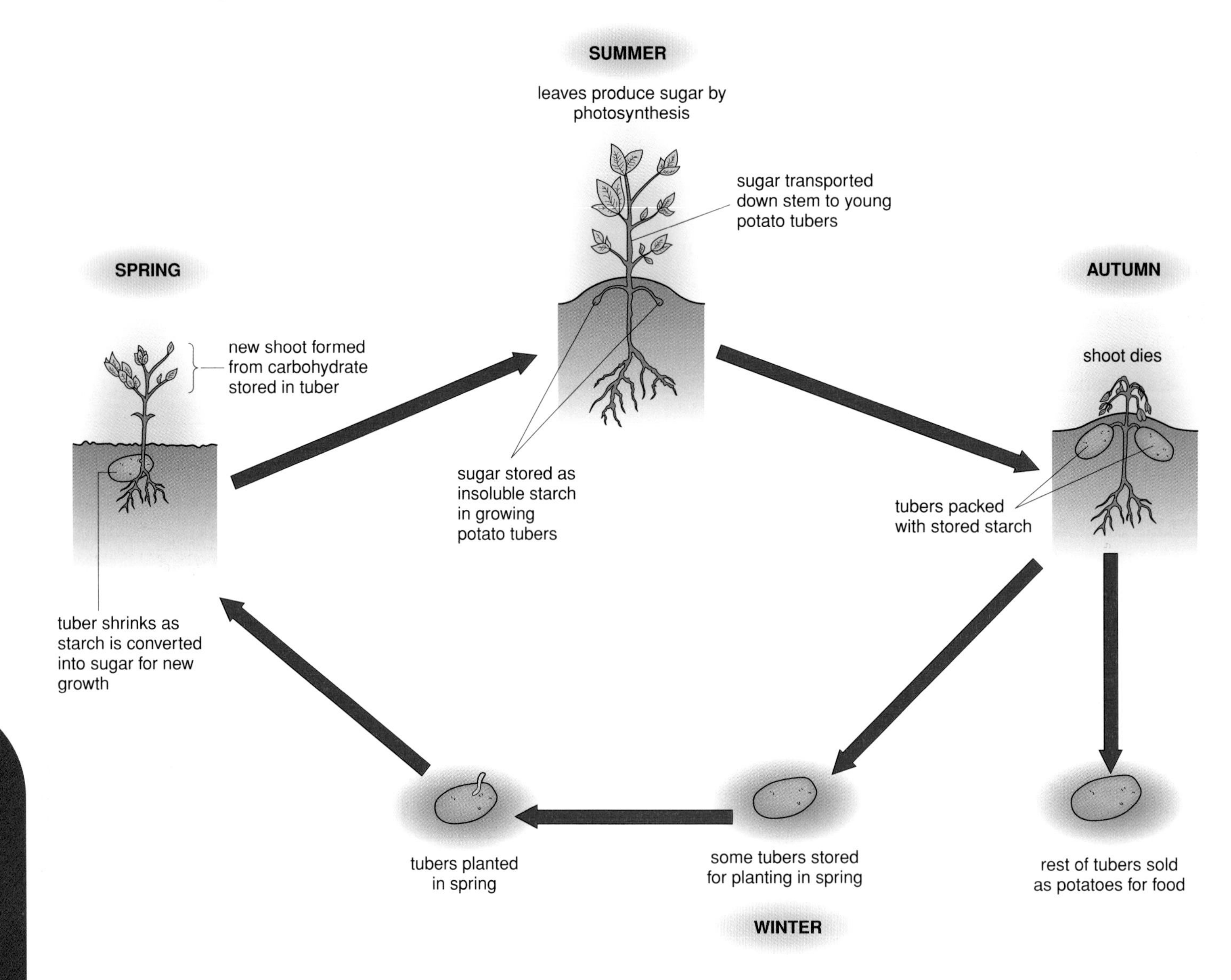

Plant respiration

All living things need energy for their life processes. Animals get their energy from the food they eat. Plants make their own food during photosynthesis, but they still need to break it down to use the energy.

Respiration in plants is very similar to respiration in animals. The big difference is that plants make their own sugar for respiration whereas animals get theirs by eating food. Respiration is breaking down glucose to release energy.

glucose + oxygen → carbon dioxide + water + [energy]

We can check if plants are carrying out respiration by detecting carbon dioxide. Plants respire 24 hours per day but we can only detect carbon dioxide being given off during the night. The diagram opposite explains why this is.

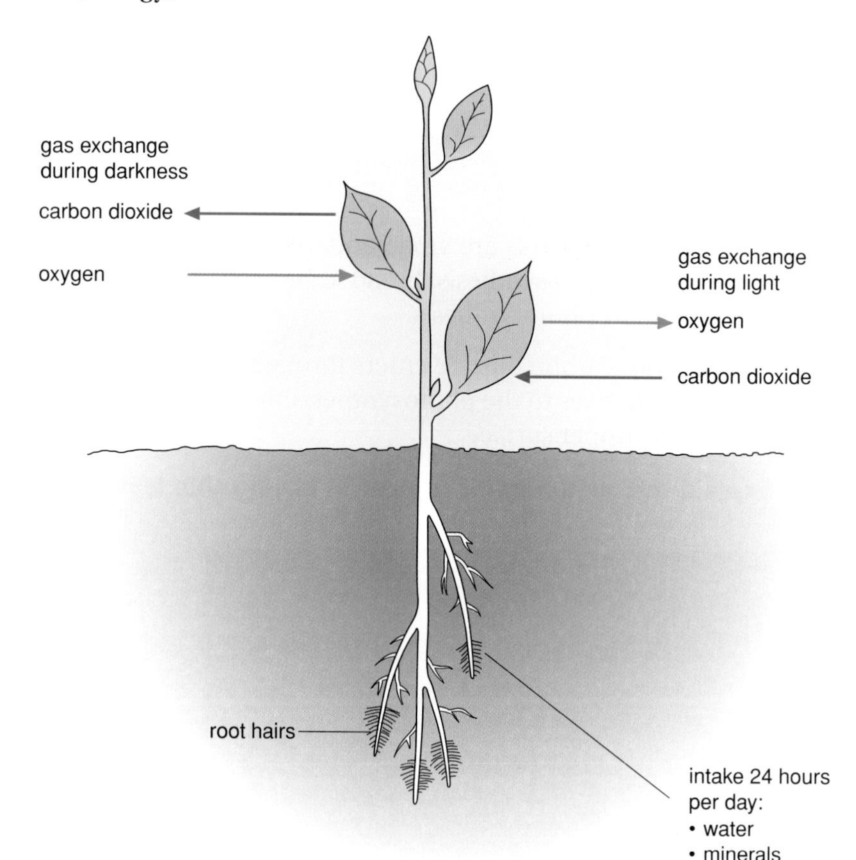

What do plants use energy from respiration for?

The sugars produced by the potato in photosynthesis are not all stored. Some are used in respiration to release energy. Most of this energy is used to make the substances needed to make new cells. New cells are needed for the tuber to grow in autumn and they are also needed to produce the shoot in spring.

Materials needed for making new cells include:

- cellulose which is needed to make the cell walls of the new cells;
- proteins: to make proteins the plant needs **nitrates** from the soil in addition to carbohydrates;
- fats.

Questions

1 Copy and complete the following sentences using the words provided:

starch   cellulose   nitrates   24   energy

Plants store carbohydrates mainly as __________. Glucose can also be converted to __________ to make cell walls. To make proteins plants need both glucose and __________

Plants respire __________ hours per day. They respire in order to release __________ from carbohydrates.

2 a Through which tissue is sugar transported from the leaves to the tubers?

b Give one important difference between sugar and starch.

c Suggest one advantage of storing carbohydrate in the tubers as starch rather than as sugar.

d Plants have similar enzymes to humans. What type of enzymes do you think plants use to convert starch to sugar in spring?

Summary

- Plant cells respire using some of the glucose produced during photosynthesis.
- The energy released by plants during respiration is used to build up smaller molecules into larger molecules:
  - sugars into starch;
  - sugars into cellulose for cell walls;
  - sugars and nitrates into amino acids which are then built up into proteins;
  - sugars into fats for storage.

2:14 Exchange surfaces

The diagram shows where some substances enter and leave plants.

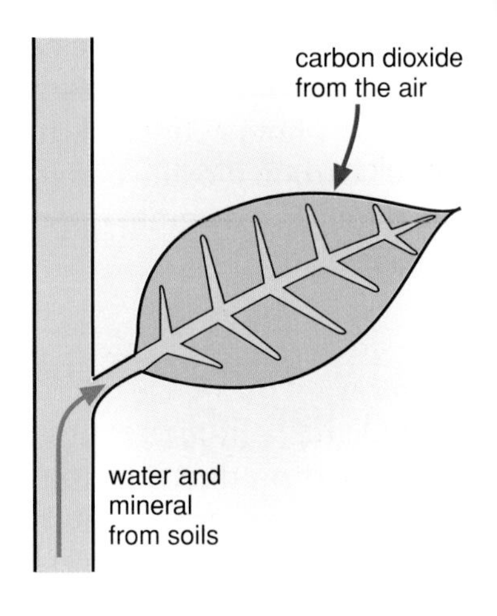

Leaves

Leaf shape

The leaves of most plants are broad and thin. Leaves absorb light energy for photosynthesis. Broad leaves will catch the most sunlight for photosynthesis.

In many leaves, carbon dioxide enters through the lower surface then diffuses to the photosynthesising cells which are just under the upper surface.

a Explain the advantage to a plant in having thin leaves.

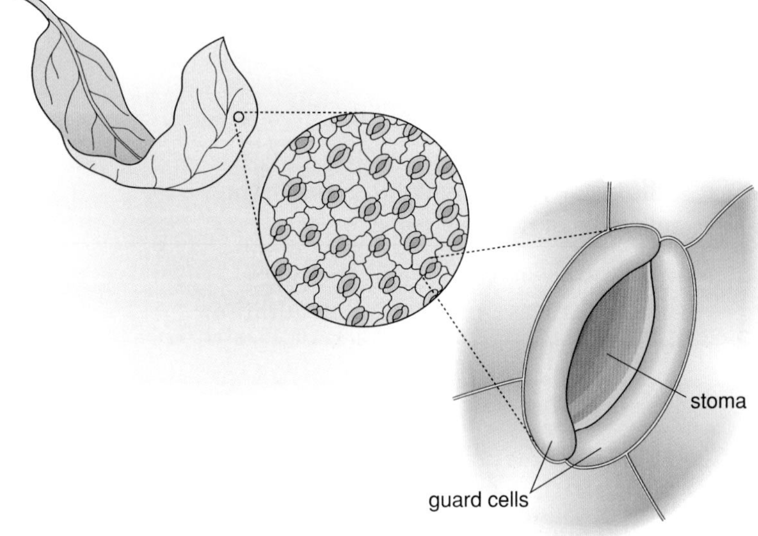

Stomata

There are thousands of **stomata** (singular, **stoma**) on each square centimetre of the leaf's lower surface. Together, they form a very large surface area through which oxygen and carbon dioxide can pass. The larger the surface area, the more gas that can pass through them.

Inside the leaf

The diagram shows how carbon dioxide gets into a leaf. Most of the chlorophyll is found in the palisade cells, near the top of the leaf and so closer to the sunlight.

Gases can diffuse in and out of leaf cells through the whole surface of each leaf cell. When added together, the total surface area of all the leaf cells is very large. This provides a very large **exchange surface**.

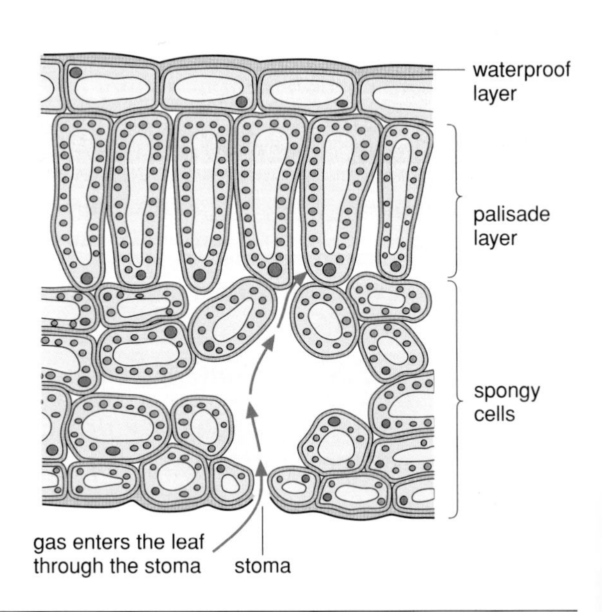

Root hairs

The diagram shows the position of **root hairs** on a root.

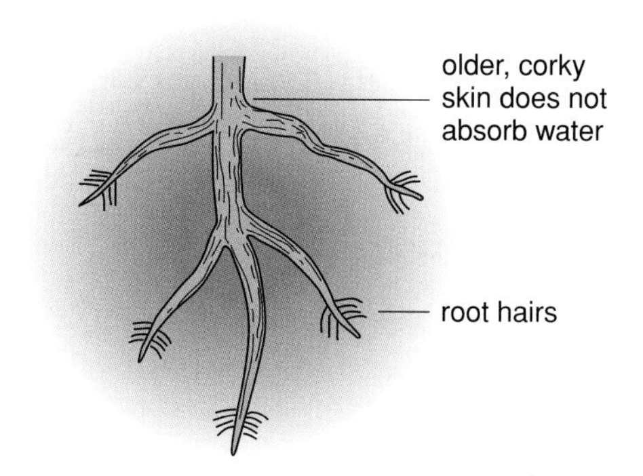

b i Describe where root hairs are situated on a root.
ii The root grows at the tip. Suggest why root hairs do not grow from the tip of the root.

The root hairs together increase the surface area of the root. This enables water and mineral ions to be absorbed more quickly.

c In humans, which structures in the small intestine increase the surface area for the absorption of food?

The soil solution is water containing low concentrations of mineral ions. This is a dilute solution. The cytoplasm and the vacuole of the root hair cell contain a higher concentration of mineral ions than the soil solution. Water moves from the soil solution into the cell by osmosis.

Questions

1 The diagram on the right shows the structure of a root hair. A root hair consists of just one cell.

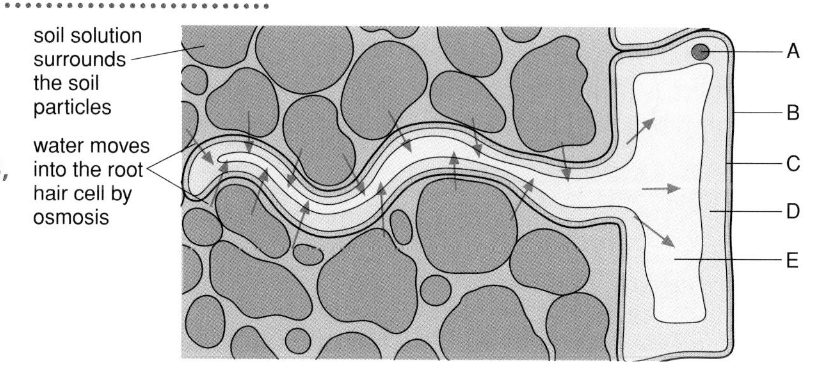

i Name the parts of the cell labelled A, B, C, D and E.
ii Explain what is meant by partially permeable.
iii Which of the structures A–E is partially permeable?

2 The photograph below shows a giant water lily. In the wild it lives in slow-moving rivers that run through jungle. Its leaves are 180 cm in diameter.

i Calculate the area of a water lily leaf.
ii What is the advantage to the water lily of having leaves of this size?

3 The photograph on the right shows a giant cactus. The cactus lives in very dry regions. To cut down water loss it has evolved leaves that have been reduced to spines. The leaf cells have no chloroplasts. The outer cells of the stem have chloroplasts and can photosynthesise.

Explain the disadvantage of the cactus's methods of photosynthesis.

Summary

Parts of plants are specialised for exchanging materials:

- The surface area of plant roots is increased by root hairs.
- The surface area of plant leaves is increased by the flattened shape and internal air spaces.

2:15 Transpiration

Evaporation

The photograph shows washing drying on a line. When does washing dry best? When it is hot, when it is windy and when it is a dry day. All of these will increase the rate at which water evaporates from the washing.

In evaporation, water vapour molecules leave the surface of the water.

- Increasing temperature increases the rate of evaporation.
- Increasing wind speed increases the rate of evaporation by moving the water vapour molecules away faster.
- Dry conditions increase the rate of evaporation because there is a lower concentration of water vapour molecules in the air so diffusion occurs more quickly.

Transpiration

Leaves normally contain a lot of water, so water will evaporate from them in exactly the same way that a wet shirt dries on a washing line.

Evaporation of water from a leaf surface is called **transpiration**. If plants lose too much water by transpiration the leaves droop. This drooping is called **wilting**. Most of the cells in the leaf are supported by being full of water.

If the vacuoles lose water, the plant cells become floppy rather than rigid. Severe wilting can kill the plant.

To reduce the amount of water being lost, most plants have a layer of wax covering the surface of the leaf. This layer of wax is called the **cuticle**.

a Cacti, which grow in deserts, have a very thick cuticle. Why do you think this is?

Opening and closing of stomata

Plants have stomata to allow carbon dioxide to enter the leaf for photosynthesis. Unfortunately for the plant, if stomata are open to allow carbon dioxide into the leaf, they will also allow water vapour to leave. The diagram shows the path taken by water vapour out of the leaf.

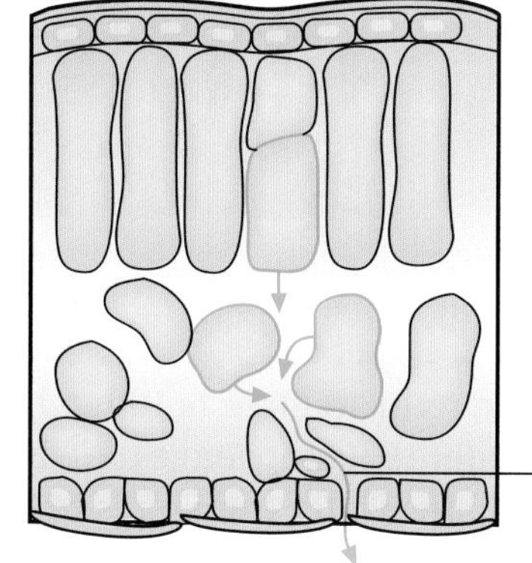

Stomata are not open all the time. The diagram below shows when stomata are open and closed during a 24-hour period.

b Explain why the 'night closing' of its stomata is an advantage to a plant.

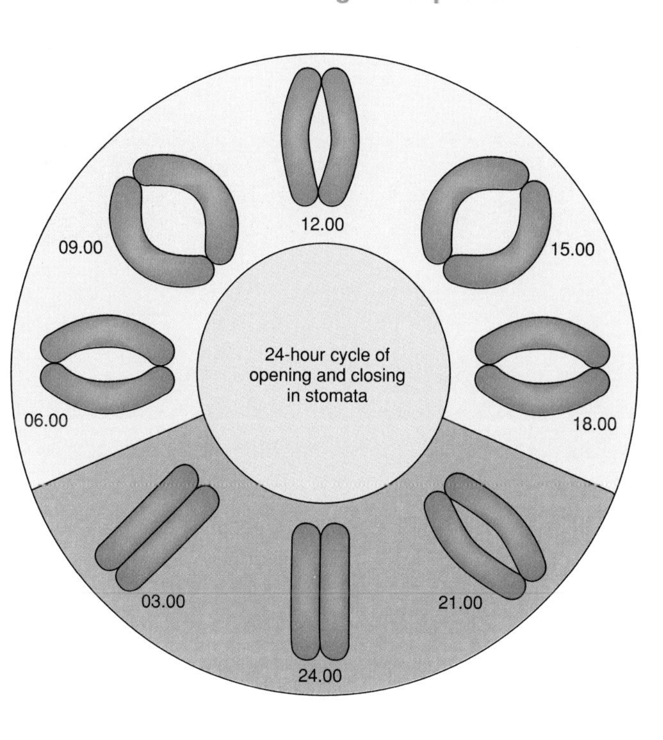

Questions

1 Copy and complete the following sentences using the words provided:

loss **evaporation**

Transpiration is the __________ of water by __________ from a plant.

2 List three conditions which increase the rate of transpiration.

3 Stomata are normally open at midday. This is when the plant is photosynthesising most rapidly. It is also the hottest part of the day, so the rate of transpiration will be the highest. If water is in short supply the plant may begin to wilt. To prevent losing too much water, the stomata may close.

How would 'lunchtime closing' of stomata affect the rate of photosynthesis?

4 In winter the water in the soil is often frozen. Many kinds of tree in this country lose their leaves in winter. Explain one advantage and one disadvantage of this behaviour to trees.

Summary

- The loss of water vapour from plant shoots is called transpiration.
- Transpiration is more rapid in hot, dry and windy conditions.
- Most plants have a waxy layer on their leaves which stops them losing too much water. Plants living in dry conditions have a thicker layer of wax.
- Plants need stomata to obtain carbon dioxide from the atmosphere.
- Because of this, transpiration occurs mainly through stomata.
- The size of the stomata is controlled by guard cells which surround them.
- If plants lose water faster than it is replaced by the roots, the stomata can be closed to prevent wilting.

2:16 Transport in plants

Xylem and phloem

Nelson's column is 56 m high. The redwood tree is 100 m high. The redwood tree needs to get water to its top leaves. We have seen how water enters a root and how water transpires from leaves. But how does water get from the roots to the leaves?

Plants have tubes for transporting materials. These tubes form two tissues in plants – **xylem** and **phloem**.

The diagram below shows where these tissues are found in a root and in a stem.

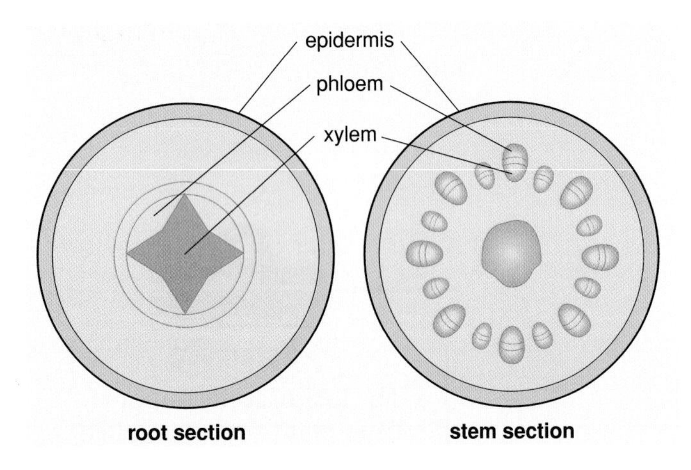

Xylem

There are several different kinds of cells in xylem, but the ones that transport most water and minerals to the leaves are called **xylem vessels**.

Xylem vessels are formed from dead xylem cells whose end walls have broken down. These form continuous tubes. The xylem vessels act like pipes in transporting water up the plant.

Wood is made up of xylem cells. Most xylem cells have substances in their walls which strengthen the cells. This is why wood is a strong material. The diagram shows how xylem vessels form.

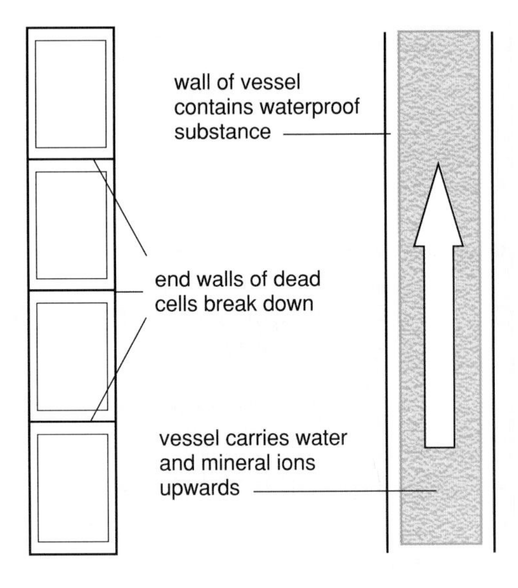

How xylem vessels form.

Phloem

There are several different kinds of cells in phloem but the ones that transport most substances are called **sieve tubes**.

The end wall of cells that form sieve tubes do not break down completely like xylem cells. Instead they form perforations. These look like a sieve. The main substance transported by phloem is sugars.

a Look back to the diagram of the potato life cycle (page 60). In which direction does phloem transport sugar:

i In summer?

ii In spring?

Movement of water through the plant

The diagram shows how water moves up through a plant.

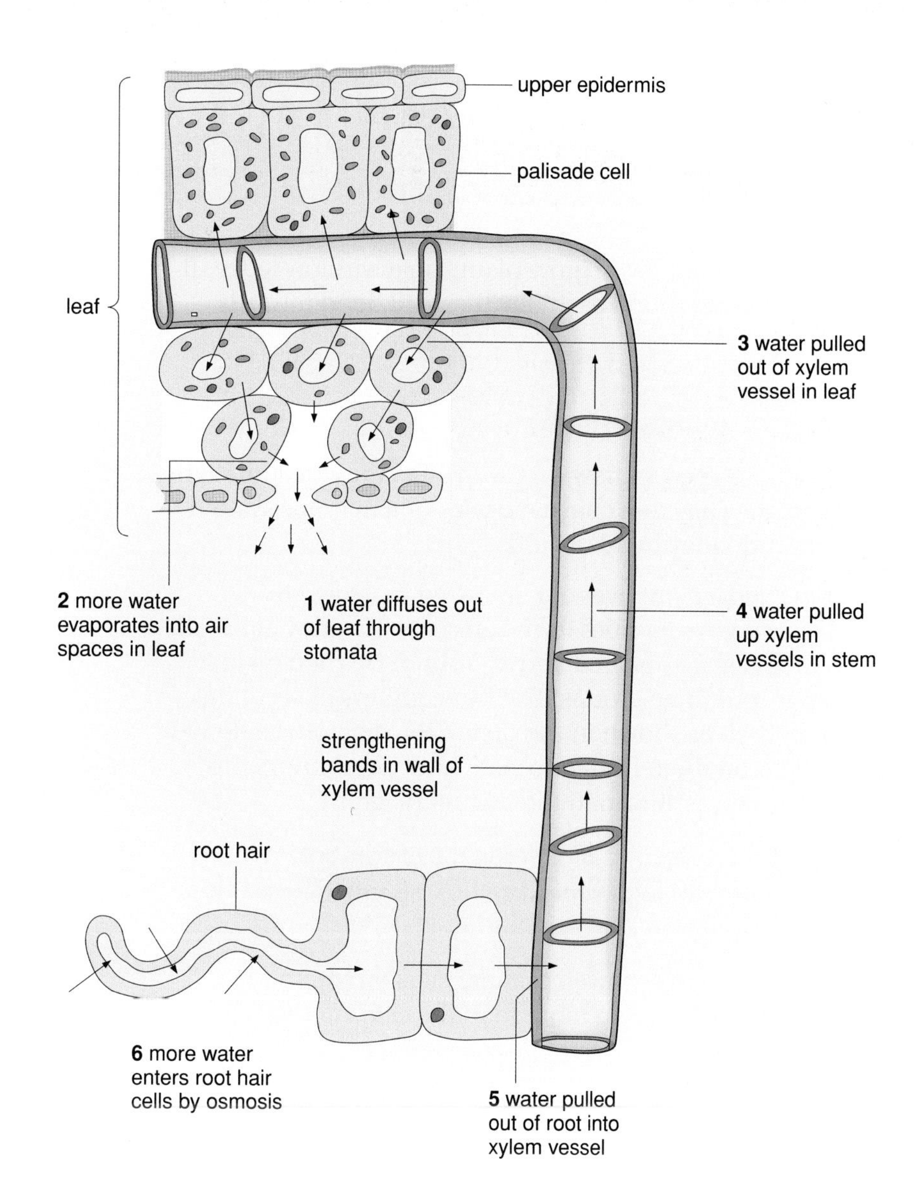

Mineral salts

Most mineral ions are transported up the stem in the water passing up the xylem.

Leaves change colour in autumn because molecules such as chlorophyll are broken down before the leaves are lost. This is so that the mineral ions they contain can be saved to produce new molecules next spring.

Questions

1 Copy and complete the following sentences using the words provided:

xylem **evaporates** **phloem** **xylem**

Water is transported up the stem through the __________ vessels. Water molecules are pulled up through the stem when water __________ from the leaves.

Sugars are transported around the plant via the __________.

Mineral ions are transported up the stem mainly via the __________.

2 Through which tissue will mineral ions be transported out of the leaf in autumn? Give the reason for your answer.

3 The rate of water movement up the stem slows down during the night. Give two reasons why.

Summary

- Water is absorbed through root hair cells by osmosis.
- Flowering plants have separate transport systems for water and nutrients:
  - xylem tissue transports water and minerals from the roots to the stem and leaves;
  - phloem tissue carries sugars from the leaves to the rest of the plants including the growing regions and the storage organs.

2:17 Plant hormones

You have to be careful if you grow plants on a window ledge. If you don't turn them round each day they will bend towards the light.

Responses of plants to light

Plants don't have eyes, so how do they detect the direction light is coming from? They don't have muscles or a nervous system either, so how do they bend towards light?

Experiments carried out over one hundred years ago by Charles Darwin gave us clues to the answers. He grew oat seeds until the shoots just appeared above ground. He then divided the seedlings into three groups. The three groups of seedlings were then treated as shown in the first of the diagrams below. He shone light on them from one side. After a few hours the shoots had grown as shown in the second diagram.

a i Describe the results of Darwin's experiment.
ii What does the experiment tell us about the part of the shoot that is sensitive to light?

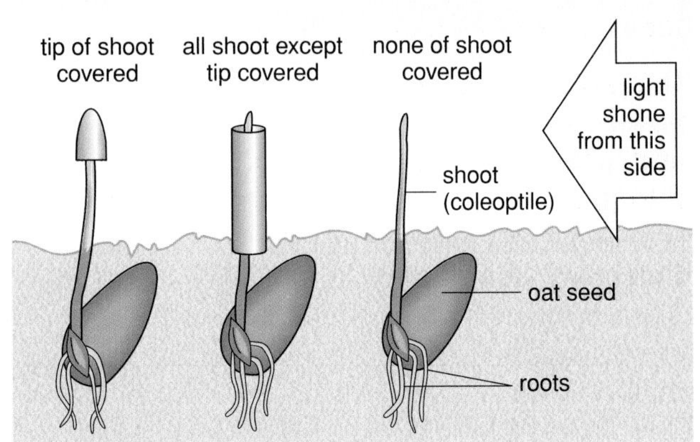

Looking at the effect of light on shoot tip growth.

Careful measurements showed that the cells on the shaded side of the shoot had grown more than the cells on the illuminated side.

In Darwin's experiment the shoot with the covered tip grew straight up and did not bend towards the light.

But how does the tip affect the growth of the part of the shoot below it?

When hormones were discovered in animals, scientists wondered if hormones controlled growth in plants. Could a hormone diffuse out of the tip and cause the cells on the shaded side to grow faster?

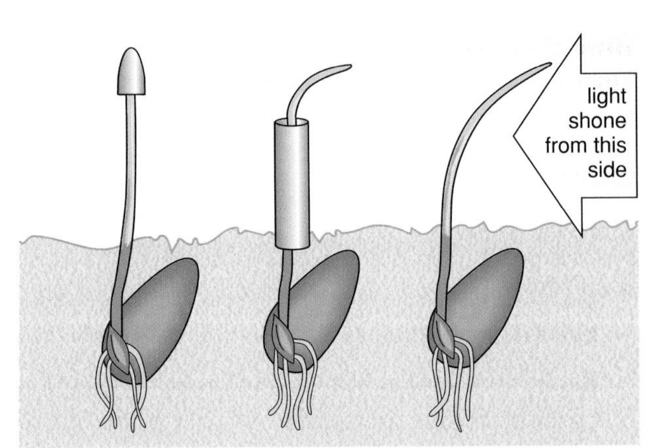

The shoots will bend towards light when the tip is exposed to light.

Plant shoots grow towards light because hormone produced by the tip diffuses more to the shaded side than the illuminated side. The cells on the shaded side grow faster. Light is the stimulus and the bending of the shoot is the response. Underground, plant roots grow away from light.

b Can you think of the advantage to the plant of

i Shoots growing towards light?
ii Roots growing away from light?

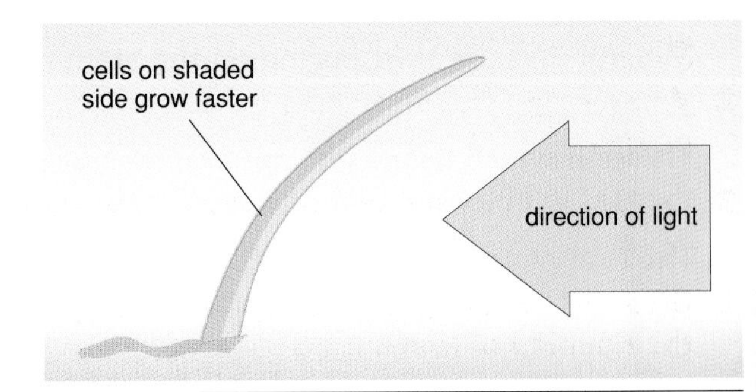

Responses of plants to gravity

Plants also respond to gravity. Plant shoots grow against the force of gravity – they grow upwards. Plant roots grow in the direction of the force of gravity.

Both of these responses are caused by gravity affecting the diffusion of hormone from the root tip into the rest of the roots.

c Can you think of the advantages of these responses to the plants?

Using hormones to control growth

Over the past hundred years, many chemicals have been used to control plant growth. Many of these have molecules with very similar shapes to natural plant hormones. These artificial hormones are used in three main ways.

Cuttings

If you cut the top of a young plant shoot and leave it in water for a few weeks it may develop roots.

Using rooting powder can speed up this process. Rooting powder stimulates roots to grow at the base of stems. The cut end of the shoot is dipped in rooting powder and then pressed into soil. After a few days roots will have grown.

Selective weed killer

A group of hormones have been found which disrupt the growth of broad-leaved plants, such as weeds. These hormones make the shoots of the plants grow much faster than the roots. The plants eventually wilt and then die. These hormones do not affect narrow leaved plants such as grasses.

Ripening hormones

Hormones have been discovered that will make fruit on the trees in an orchard all ripen at the same time.

Summary

- Plant shoots grow towards light and against the force of gravity.
- Plant roots grow towards moisture and in the direction of the force of gravity.
- These responses are brought about by changes in the distribution of the hormones that bring about plant growth.
- Plant hormones are used by humans to:
  - produce large numbers of plants quickly by stimulating the growth of roots in cuttings;
  - regulate the ripening of fruits on the plant and during transport to consumers;
  - kill weeds by disrupting their normal growth patterns.

Questions

1 Copy and complete the following sentences using the words provided:

light   hormones   moisture   gravity

Plant stems grow towards __________. This response is caused by chemicals called __________ moving to the shaded side of the stem. Plant roots also grow towards __________ and in the direction of __________.

2 What is the advantage to a grower of using hormone rooting powder?

3 Explain why weeds wilt after they have been sprayed with selective weed killer.

4 Give the advantages to a farmer of using hormones that affect the ripening of fruits.

End of module questions

1 The drawing shows a cell from the leaf of a plant.

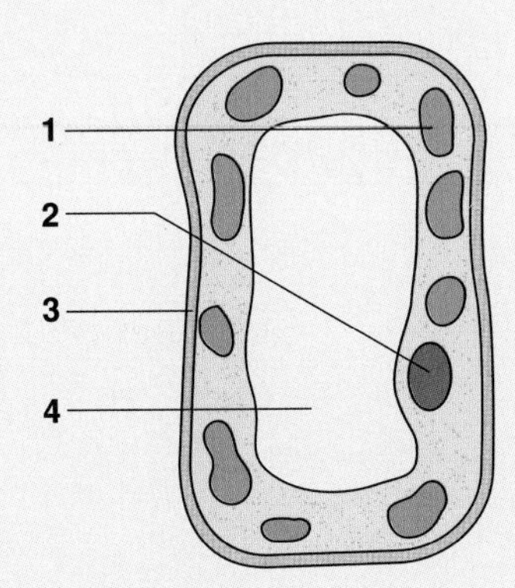

Match words from the list with each of the labels 1–4 in the diagram.

cell wall
chloroplast
nucleus
vacuole

2 The table has information about different receptors in the body of a lizard which lives in a hot, rocky place.

Part of the body	Contains receptors which allow the body to
1	feel how hot the rocks are
2	see its prey
3	hear other animals approaching
4	taste its food

Match words from the list with each of the numbers 1–4 in the table.

ear
eye
tongue
skin

3 The diagram shows the parts of a plant.

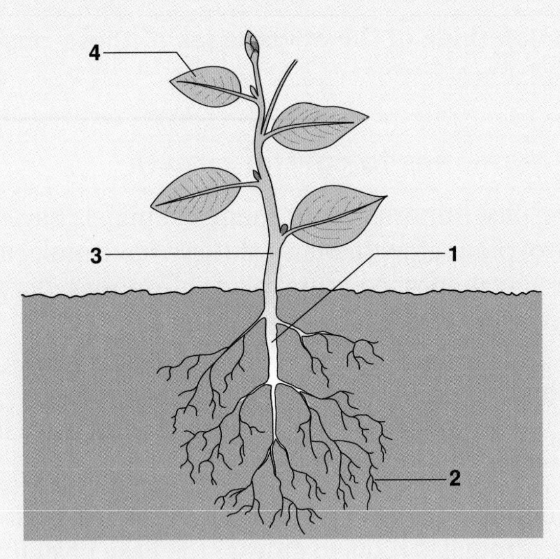

Match words from the list with each of the labels 1–4 in the diagram.

absorbs mineral ions
holds the leaves upright
makes food
transports water to the stem

4 The table is about the growth of plants which may be affected by a number of factors.

Factor	Plant activity
1	roots grow in the direction of it
2	stimulates the ripening of fruits
3	shoots grow towards it
4	absorbed by roots

Match words from the list with each of the numbers 1–4 in the table.

hormone
light
moisture
the force of gravity

5 The drawing shows a section through a plant leaf.

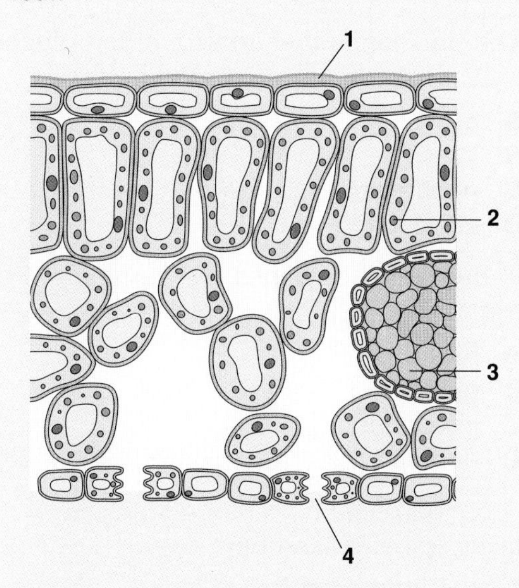

Match words from the list with each of the labels 1–4 in the drawing.

where carbon dioxide enters the leaf
carries sugars to the roots
reduces evaporation of water
where photosynthesis takes place

6 Smoking may harm the body. Which **two** of the following are most likely to be damaged by tobacco smoke?

brain
blood vessels
heart
liver
kidneys

7 Plants make food by photosynthesis. Which two of the following are raw materials for photosynthesis?

carbon dioxide
glucose
oxygen
starch
water

8 The diagram shows a section through the eye.

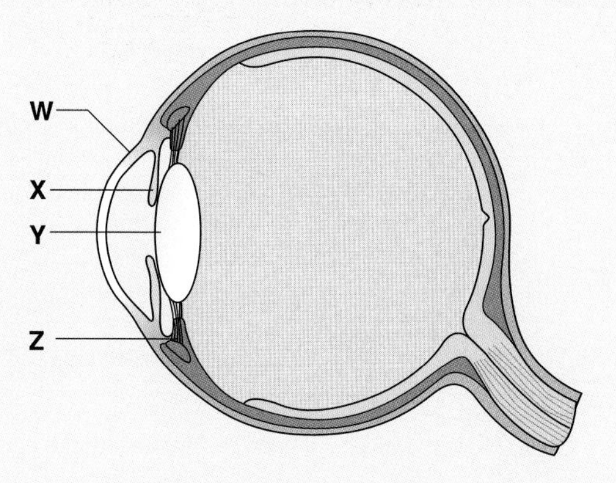

8.1 Which parts produce the image on the retina?

A W and X
B W and Y
C X and Y
D W and Z

8.2 Which part is transparent?

A cornea
B sclera
C optic nerve
D iris

8.3 Which part contains receptors sensitive to light?

A sclera
B iris
C optic nerve
D retina

8.4 Information is passed from the eye to the brain by

A hormones
B the blood system
C sensory neurones
D motor neurones

9 The drawing shows the main daily input and output of water for an adult.

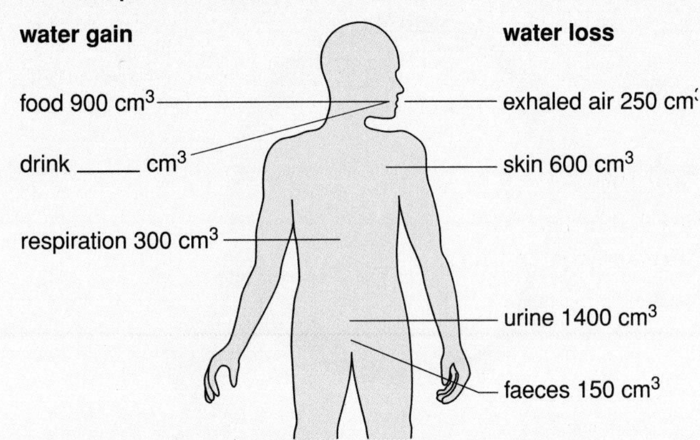

9.1 How much water did the person drink in one day?

A 400 cm^3
B 1000 cm^3
C 1400 cm^3
D 2400 cm^3

9.2 What proportion of the water loss was via the skin?

A 1/4
B 1/10
C 1/40
D 1/400

9.3 Which organ produces urine?

A kidney
B liver
C bladder
D lungs

9.4 Which organ stores urine?

A kidney
B liver
C bladder
D lungs

10 The body needs to control the concentrations of substances in the blood.

10.1 The hormones that control blood glucose concentration are produced by the

A brain
B liver
C pancreas
D stomach

10.3 In humans, hormones are transported by the

A blood
B glands
C intestines
D nerves

10.3 A dangerously high blood sugar level is likely to be caused by

A too little exercise
B eating too much food
C sweating a lot
D insufficient insulin

10.4 If we have too many amino acids the excess are broken down

A into urine in the bladder
B into urea in the bladder
C into urea in the liver
D into urea in the kidney

Module 5 – Metals

If you travel by bus, train, bike or car you will know how important metals are. Iron and steel are used to build the great machines and structures that are all around us. And don't forget that there are many other metals. You also rely on copper wires to carry electricity to your home and school, and aluminium for your drink cans and saucepans. Aluminium is even used to build aeroplanes!

You already know about the basic properties of metals and how they react with oxygen and acids. In this unit you will find out more about the chemistry of metals:

- how they are arranged in the periodic table;
- how the alkali metals, such as sodium, differ from the transition metals, such as iron or copper;
- how the reactivity of metals affected their discovery;
- how iron and aluminium are made from their ores;
- how you can stop iron from rusting;
- how useful chemical compounds called salts can be made from metals.

Before you start, check what you remember about metals and their reactions.

1. How can you tell a metal from a non-metal?
2. What compounds form when metals react with air?
3. What happens when some reactive metals are put into acid?
4. What causes iron to rust?
5. What is the pH scale and how is it used?

5:1 The periodic table

What a lot of metals

Scientists have arranged the **elements** into a pattern of rows and columns called the **periodic table**. In this table, elements with similar properties are found in the same column. These columns are called **groups**.

The periodic table is a very useful way of looking at the elements. You can tell a lot about an element just by its position in the table.

All the elements in group 1 are very reactive metals. All the elements in group 7 are reactive non-metals.

There are many more metals (shaded red) than non-metals (shaded green) in the periodic table. More than three-quarters of the elements are metals.

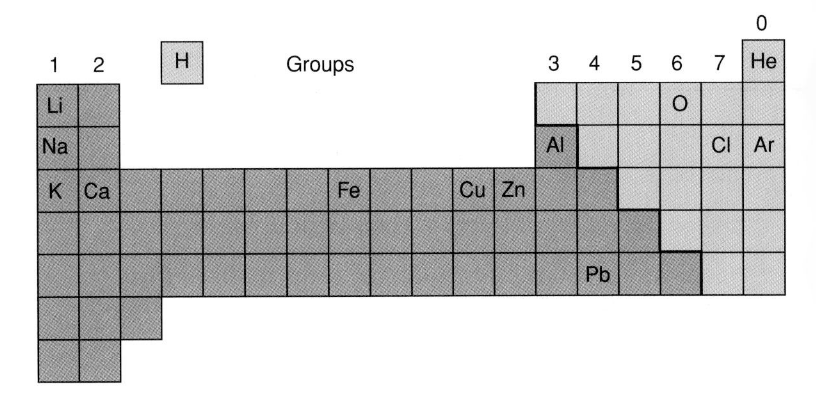

There is a full version of the periodic table in the data sheet section.

a Which symbols represent the elements calcium, zinc, aluminium and oxygen?

b Helium (He) is an unreactive gas. What do you think argon (Ar) is?

c The old name for iron was *ferrum*. The symbol for iron is shown in the middle of the periodic table. What is it?

So what are metals anyway?

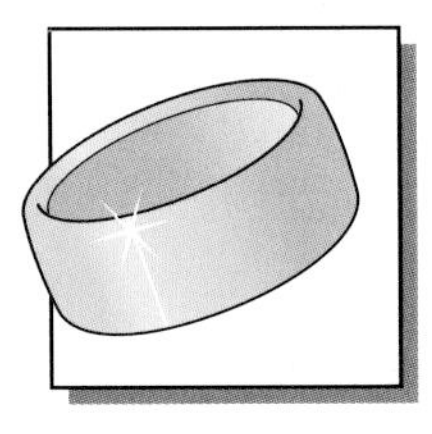

Metals are shiny.
Many are silvery, though some are yellow or brown.

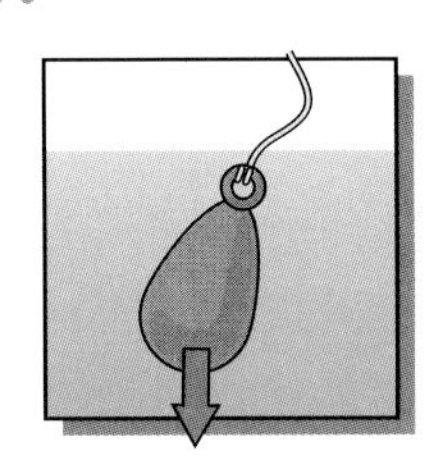

Most metals are dense.
Iron is nearly eight times as dense as water.

Most metals are solids.
The exception is mercury.

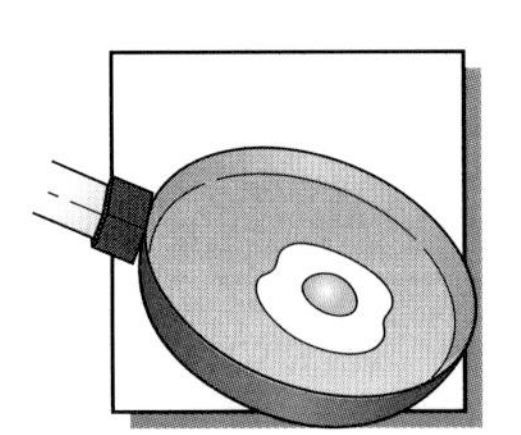

Metals conduct heat.
That's how heat energy gets to the eggs in a frying pan.

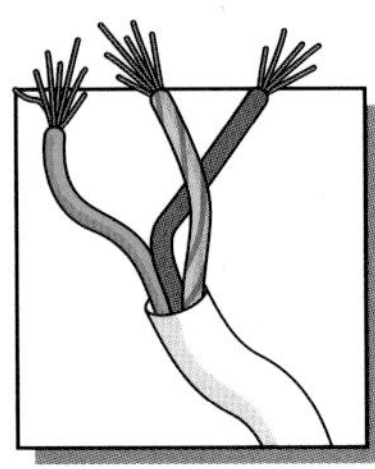

Metals conduct electricity.
Copper is used for electric wiring.

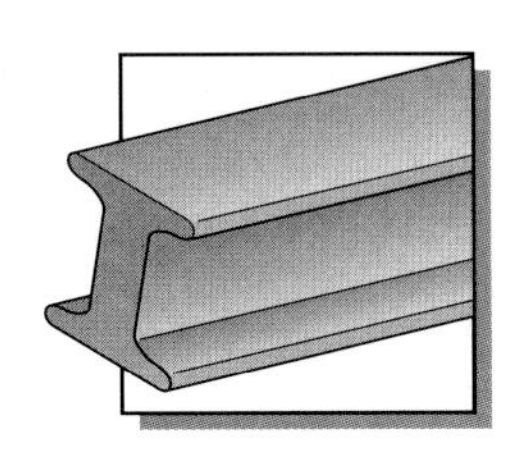

Many metals are hard and strong.
Steel is used for bridges and machines.

How is the periodic table arranged?

The first periodic table was made by Dimitri Mendeleev. He put the elements in order of the 'heaviness' of their atoms. This is called **relative atomic mass**.

All atoms are made from small particles called **protons**, **neutrons** and **electrons**. Larger atoms have more mass.

Mendeleev found that his periodic table gave him groups of elements with similar properties, but when new elements were discovered some of them did not fit the pattern. For example, argon (which is a very unreactive gas) appeared in group 1 with the very reactive metals. Potassium (a very reactive metal) appeared in group 0 with the unreactive gases.

The periodic table we use today arranges atoms by their **atomic number** (or **proton number**). This is the number of **protons** in the **nucleus** their atoms.

Dimitri Mendeleev.

Questions

1 Copy out and complete these sentences. Choose from:

left **group** **metals** **relative atomic** **periodic**

In the __________ table, the elements are arranged in order of increasing __________ mass. Similar elements occur in the same vertical __________. Most elements are __________. They are found in the middle and on the __________ of the table.

2 a Arrange these metals in order of increasing melting point and plot a bar chart to show this.

Metal	Melting point (°C)
aluminium (Al)	660
copper (Cu)	1080
iron (Fe)	1540
lead (Pb)	327
mercury (Hg)	–39

b The temperature drops to –50 °C in Siberia in winter. Why can't you use a mercury thermometer to measure this?

3 Copy out two correct sentences by choosing the correct phrase from each pair.

If potassium and argon are arranged in order of their (relative atomic mass/proton number) they appear in the (correct/wrong) groups.

4 Mendeleev's periodic table worked well at the time. Why did scientists need to revise it later?

Summary

- The periodic table arranges the elements in family groups.
- The elements are arranged in order of increasing atomic mass.
- Most elements are metals. They are found in the middle and on the left of the periodic table.

5:2 The alkali metals

Group 1

Group 1 of the periodic table contains a family of very reactive metals. They include lithium (Li), sodium (Na) and potassium (K).

a Which metals had the old-fashioned names *natrium* and *kalium*?

Like all metals, the metals in group 1 are:

- good conductors of heat;
- good conductors of electricity;
- very shiny when fresh (though they tarnish in air).

But, unlike most 'everyday' metals, they:

- are very soft and can be easily cut with a knife;
- have low densities and these three can float on water;
- have relatively low melting and boiling points.

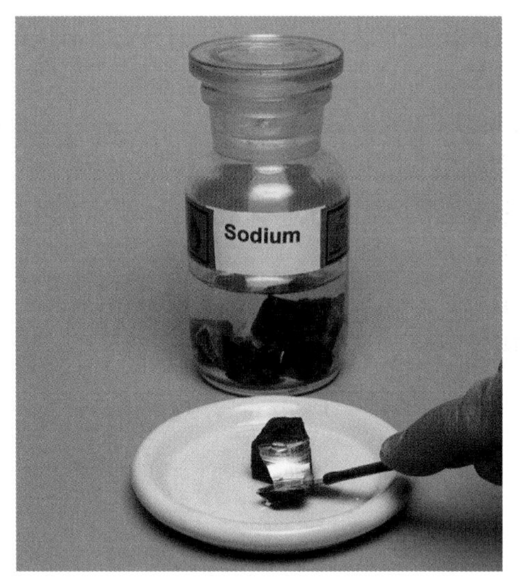

Sodium is a typical group 1 metal – very reactive and very soft.

Burning in air

The group 1 metals are very reactive. They are stored under oil to stop them reacting with air or water.

Group 1 metals react with oxygen in the air to form the metal oxide. The oxide of a metal is a powdery compound.

When lithium burns in air you see a bright red flame.

lithium + oxygen → lithium oxide

b Why do you think lithium compounds are sometimes used in fireworks?

Lithium burns in air with a bright red flame.

The alkali metals

Group 1 metals react with water. They form a metal hydroxide. Hydrogen gas is also given off in this reaction.

sodium + water → sodium hydroxide + hydrogen

Sodium hydroxide dissolves in water. The solution is a strong **alkali**. All group 1 metal hydroxides make strongly alkaline solutions in water. Because of this, group 1 metals are called the **alkali metals**.

Sodium reacts violently with water.

Common salt

The alkali metals react with non-metals to form white compounds. These dissolve in water to give colourless solutions. Sodium reacts with chlorine to give sodium chloride – common salt.

These compounds *are* different even though they all look similar!

Introducing ions

In some chemical reactions, atoms gain an electric charge. Particles that have gained an electric charge are called **ions**.

When alkali metals react, they turn into ions. Alkali metal ions carry a positive charge.

Compounds such as sodium chloride are made from ions. They are called **ionic compounds**.

c What is the difference between a sodium atom and a sodium ion?

What's the trend?

d Draw a bar chart to show the melting points of these three metals.

Metal	Melting point (°C)	Reaction with water
lithium	180	fizzes steadily
sodium	98	very rapid reaction
potassium	64	violent reaction

e What pattern does this show as you go down group 1?

f The next element in the group is rubidium. Will its melting point be higher or lower than that of potassium?

g What pattern do you see in the way the metals react with water as you go down group 1?

h Do you think it would be safe to put rubidium in water? Explain your answer.

Questions

1 Copy and complete these sentences. Choose from:

hydrogen **water** **reactive** **alkaline**

The alkali metals are a family of soft, __________ metals. They react with __________, forming a strongly __________ solution. The gas given off is __________.

2 List three ways in which sodium is a typical metal and three in which it is unusual.

3 Choose from the paired words to make a correct sentence:

Sodium forms a (positive/negative) (atom/ion) when it reacts to form sodium chloride.

Summary

- Group 1 metals are soft and have a low density.
- Group 1 metals are very reactive. They form the metal hydroxide and hydrogen gas with water.
- Group 1 metals react with non-metals to form white compounds such as common salt.

5:3 The transition metals

The 'everyday' metals

The block in the middle of the periodic table contains the familiar, 'everyday' metals such as iron and copper. These metallic elements are called the **transition metals**.

Like the alkali metals, the transition metals are shiny, and are good conductors of heat and electricity.

Unlike the alkali metals they:

- are hard, tough and strong;
- have high melting points (except mercury).

Copper

Copper has some very useful properties:

- it is a particularly good conductor of electricity;
- it is easy to bend and shape;
- it does not react (**corrode**) much with water and air.

Iron and steel

Steel is a very useful form of iron. Iron and steel have even more uses than copper. Here are their properties:

- they are hard, tough and strong;
- they can be hammered and bent into any shape when hot;
- they have very high melting points.

Unfortunately, iron and steel corrode in air and water – they rust.

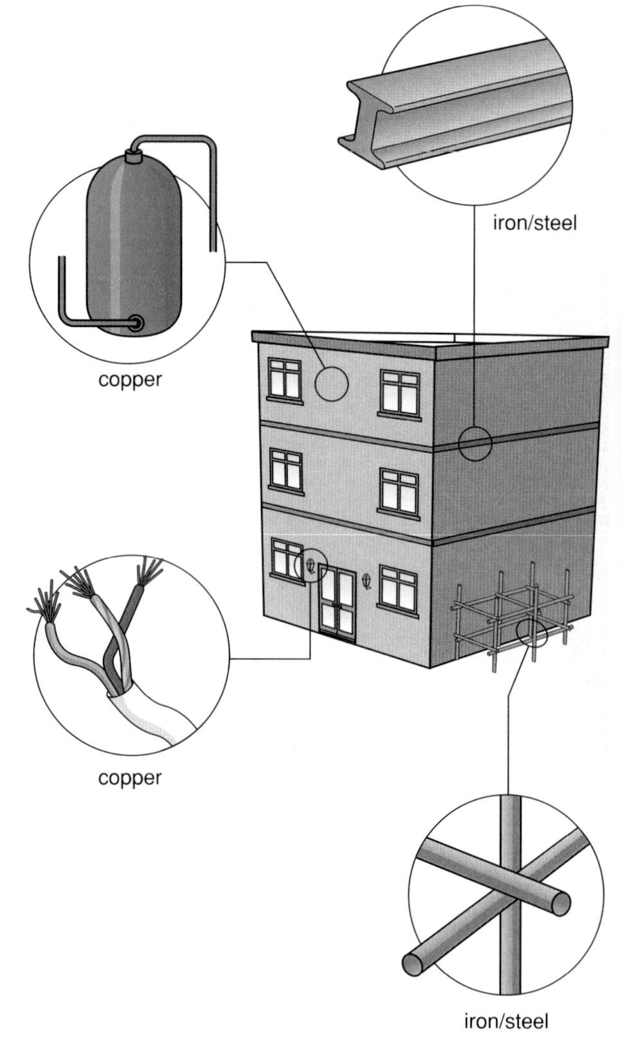

Uses of iron, steel and copper.

a What property of iron and steel causes problems in time? (Think about what happens to old cars ...)

b List three uses of iron and three of copper.

c Copy and complete this table, showing which properties of copper are useful for which use.

Use of copper	First property	Second property
electric wires		
water pipes		

d Copy and complete this table, showing which properties of steel are useful for which use.

Use of steel	First property	Second property
girders for bridges		
making car engines		

Transition metal compounds

Transition metals react to form compounds. Unlike the alkali metals, transition metal compounds are often brightly coloured.

You are probably familiar with the blue colour of copper sulphate. Many copper compounds are blue, but others are green. When copper roofs weather they turn a beautiful green colour.

e What metal has been used for the roof of this building?

Some iron compounds are green, but others are rusty brown. Rust is in fact iron oxide. Both iron and copper compounds are used to colour the glazes on pottery.

f Why are transition metal compounds used as glazes on pottery?

Did you know?

The properties of transition metals can often be improved by mixing them up. These mixtures are called **alloys**.

Pure gold is too soft to use for jewellery. It has to be mixed with copper to make it hard enough to use.

Questions

1 Copy and complete the following sentences. Choose from:

electricity **transition** **melting** **copper** **tough**

The __________ metals are a family of useful metals including __________ and iron. Like the alkali metals they conduct heat and __________ well. Unlike the alkali metals they are hard and __________ and have high __________ points.

2 Ahmed had some iron sulphate and copper sulphate crystals in two unlabelled jars. How could he tell which was which?

3 Manganese forms pink or purple compounds. Is manganese an alkali metal or a transition metal?

4 Gold is an excellent conductor of electricity. Why isn't gold used instead of copper for house wiring?

Summary

- The common metals iron and copper belong to the transition metal family.
- Transition metals are harder, stronger and less reactive than the alkali metals.
- Transition metal compounds are often coloured.

5:4 Metals from the Earth

Metals in the Earth

The rocks of the Earth contain all the metals we need – but they are not usually easy to get at!

- Some metals are common but others are rare.
- Most are locked up in chemical compounds.
- These compounds are all mixed up in the rocks.

Fortunately, metal compounds are sometimes concentrated by natural processes. Rocks like this are called **ores**. It is easier and cheaper to get at the metal compounds when they are concentrated like this.

From left to right, these ores are malachite, a copper ore, galena, a lead ore, and pyrites, which could be used as an iron ore.

Metal ores

It is not always easy to get the pure metal away from its compounds. The more reactive the metal, the stronger its compounds will be. The stronger the compound, the harder it is to get the metal out! Aluminium and iron are common, but they are so reactive that they are normally tightly locked up in compounds.

Gold is so unreactive that it is found as the pure metal. It may be rare but, if you were very lucky, you might find a pure, glistening lump of it.

How do they react with air?

Most metals will react with the oxygen in the air to form the metal oxide. For example:

magnesium + oxygen → magnesium oxide

But the way in which the metal reacts varies considerably:

- strips of magnesium react so quickly they burn easily;
- iron will only burn as a fine powder – that's how 'sparklers' work;
- copper does not burn, but will react slowly;
- gold does not react at all.

a **Write a word equation for iron burning in oxygen.**

b **Arrange copper, gold, iron and magnesium in order of increasing reactivity with oxygen.**

Magnesium burns with a brilliant white flame.

How do they react with water?

Many metals react with water to give the metal hydroxide. For example:

sodium + water → sodium hydroxide + hydrogen

But again, the way they react varies:

- sodium reacts violently;
- calcium fizzes in cold water;
- magnesium will react with hot water or steam;
- other metals react much more slowly or not at all.

c Arrange calcium, iron, magnesium and sodium in order of increasing reactivity with water.

How different metals behave in hydrochloric acid. (For question d, below.)

How do they react with acids?

Many metals react with acid, giving off hydrogen gas. For example:

hydrochloric acid + magnesium → magnesium chloride + hydrogen

Again, the way they react varies:

- potassium and sodium react so violently with acid that this is far too dangerous to try in school!
- calcium and magnesium react very vigorously;
- iron and zinc react more slowly;
- lead and tin will only react with warm, strong acid;
- copper and gold will not react in this way at all.

d Four metals are shown in the photograph above. Arrange them in order of increasing reactivity with acid.

very reactive	potassium	strong compounds
	sodium	
	calcium	
	magnesium	
	aluminium	
	zinc	
	iron	
	tin	
	lead	
	copper	
unreactive	gold	weak compounds

The reactivity series

So, different metals react at different rates. You can list them in order of reactivity, from most reactive down to the least. The order of reactivity is the same, no matter what the reaction. This list is called the **reactivity series** of metals.

Questions

1 Copy and complete these sentences. Choose from:

concentrated compounds metals mixed ores

Most __________ are found as compounds, __________ up in the rocks. Sometimes they are __________ in __________. The __________ have to be broken up to get the metal out.

2 Choose from each pair of words to make two correct sentences.

(Reactive/Unreactive) metals form (weak/strong) compounds that are (hard/easy) to break.

Summary

- Metals are found as ores in the rocks of the Earth.
- Metals can be arranged in a reactivity series based on their reactions with air, water and acid.

5:5 Displacement

Getting the metals out

Most metals are found locked up in chemical compounds, such as the metal oxide. These need to be broken apart to get the metal out. This cannot usually be done by heat alone. A suitable chemical reaction is needed.

Iron forces the copper out of copper sulphate solution.

a You *can* break up mercury compounds by heating. Where do you think you would find mercury in the reactivity series?

Displacement reactions

If you dip an iron nail into blue copper sulphate solution, you will find that it comes out an orangey-brown colour – it has been copper-plated! Some of the iron atoms have swapped places with the copper in solution.

metallic iron + copper in solution → iron in solution + metallic copper

This is called a **displacement reaction**. It happens because iron is more reactive than copper and so displaces it from solution.

Displacement and reactivity

A less 'scientific' view is that iron is a 'chemical bully', and pushes weaker copper from its compound. Whichever way you choose to think of it, the rule is that a more reactive metal will be able to displace a less reactive one.

b Which metal could bully iron out of its compound? Magnesium or lead?

A spectacular displacement

Displacement reactions can take place between solids, too. Aluminium is more reactive than iron. If you mix aluminium with iron oxide and give it a kick-start of energy, the aluminium pushes the iron from its oxide. This is called a **thermit** reaction, and it can be very spectacular, with sparks flying and molten iron left behind!

aluminium + iron oxide → aluminium oxide + iron

c Aluminium is more expensive than iron. Why isn't the thermit reaction used to make iron commercially?

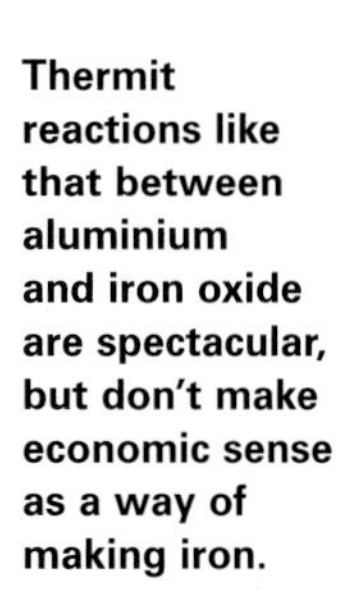

Thermit reactions like that between aluminium and iron oxide are spectacular, but don't make economic sense as a way of making iron.

The catch ...

Displacement is a good way to get metals from their compounds, but there is a major catch. You need to use a more reactive metal to do it, but the more reactive metal will be harder to get out of its compounds in the first place!

Fortunately, non-metals fit into the reactivity series too.

All acids contain hydrogen. When metals fizz in acid, they are displacing the hydrogen from it and taking its place. Metals that react like this with acid must be more reactive than hydrogen. Copper never reacts in this way because it is less reactive than hydrogen.

very reactive
potassium
sodium
calcium
magnesium
aluminium
carbon
zinc
iron
tin
lead
hydrogen
copper
gold
unreactive

d Use the reactivity series to list three metals that can displace hydrogen from acids.

Hydrogen can displace copper. If you pass hydrogen over heated copper oxide, the copper is displaced from its compound.

$$\text{copper oxide} + \text{hydrogen} \rightarrow \text{copper metal} + \text{hydrogen oxide}$$

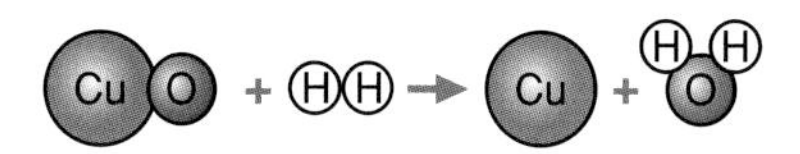

e What do we usually call hydrogen oxide?

f What is the formula of copper oxide, as shown here?

Carbon to the rescue!

Even hydrogen is expensive to produce. What is needed is a cheap, reactive element, something you could just dig up out of the ground ...

The answer to the problem is carbon. Carbon fits into the reactivity series between aluminium and zinc and so is quite capable of pushing lead, copper or iron from their compounds in a displacement reaction. It is also in plentiful supply in a fairly pure form as coal.

Carbon, as a piece of coal.

g Could you use carbon to get aluminium from aluminium oxide?

Questions

1 Copy out and complete the following. Choose from:

reactive oxide iron
Aluminium displacement

__________ can act as a chemical bully and push a less __________ metal such as __________ from its __________ ore. This type of reaction is called a __________ reaction.

2 In industry, why is carbon rather than aluminium used to displace iron from its ore?

3 List four metals that could be got from their ores using carbon.

Summary

- A more reactive element will displace a less reactive element from its compounds.
- Carbon can be used to displace some less reactive metals from their ores.

5:6 Making iron

The iron age

Iron can be displaced from its ores by carbon, but very high temperatures are needed to kick-start this reaction. Because of this, iron was not discovered until long after copper and lead. Today, more iron is used than all the other metals together.

Metal	Amount produced (million tonnes per year)
aluminium	12.5
copper	8
iron	400
lead	3
tin	0.2
zinc	6

a Arrange the metals in increasing production tonnage.

b Approximately what percentage of the total is iron? (Choose from 9%, 19%, 49% or 90%.)

Reducing iron

In simple terms, iron is made from iron oxide ore by a displacement reaction with carbon. The carbon 'steals' the oxygen away from the iron.

When oxygen is taken away like this, the process is called **reduction**. The production of iron in this way is often called **carbon reduction**.

$$\text{iron oxide + carbon} \xrightarrow{\text{reduction}} \text{iron + carbon dioxide}$$

c Lead oxide can be reduced to lead by carbon. Write a word equation for this reaction.

The raw materials

Iron is produced by carbon reduction in a **blast furnace**. For this, the following raw materials are needed.

- Iron ore. The commonest iron ore is haematite (Fe_2O_3).
- Coke. Carbon is needed to burn in the furnace to make it hot. It is also needed to reduce the iron oxide ore. Modern furnaces use coke. This is coal which has been heated to drive off oils and gases, leaving fairly pure carbon (C).
- Limestone. This is calcium carbonate ($CaCO_3$). It is added to remove acidic impurities.
- Oxygen is also needed for the carbon to burn. This is provided by blasting hot air into the furnace.

d Why is this furnace called a *blast* furnace?

A blast furnace in operation.

How it works

Iron is produced in the blast furnace by a three-stage reaction.

- **Stage 1.** The coke burns, giving off lots of heat.

  carbon + oxygen → carbon dioxide

- **Stage 2.** The carbon dioxide reacts with more carbon.

  carbon dioxide + carbon → carbon monoxide

- **Stage 3.** The carbon monoxide reduces the iron oxide.

  carbon monoxide + iron oxide → carbon dioxide + iron

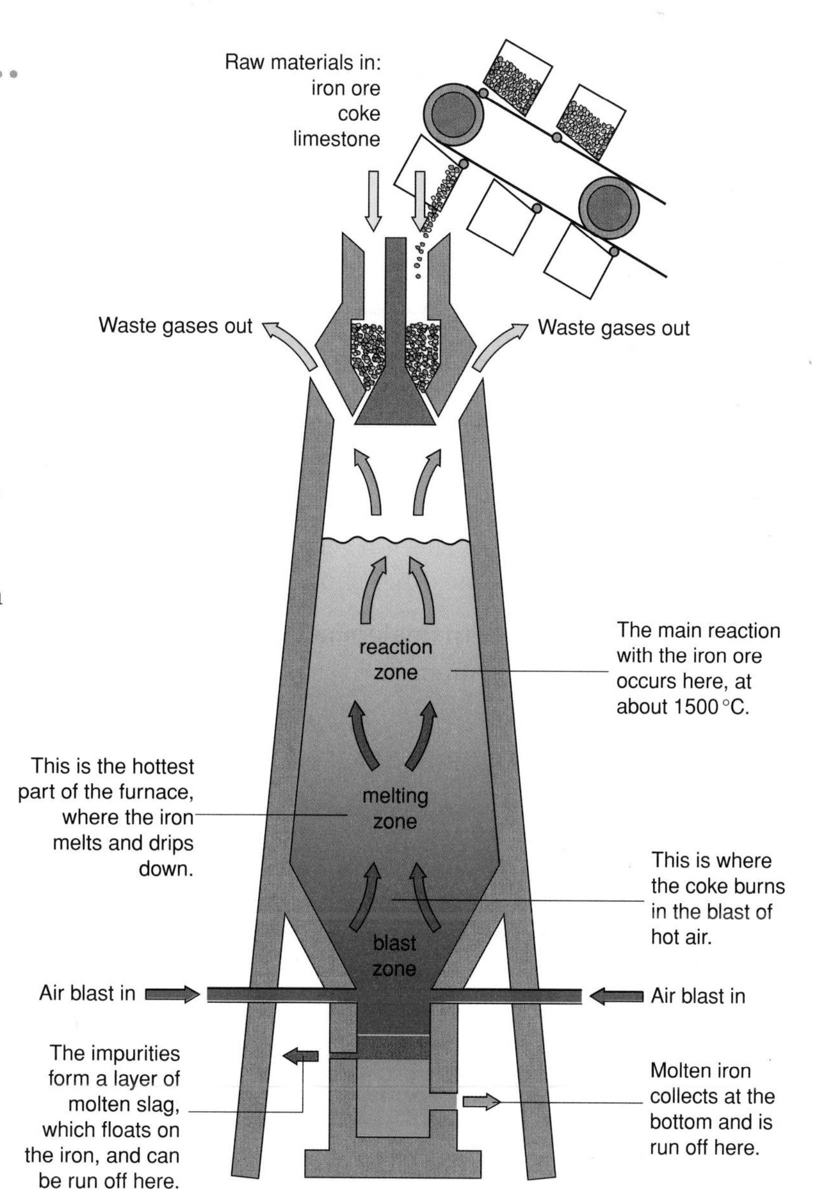

Questions

1 **Copy and complete the following sentences. Choose from:**

reduces dioxide carbon iron

Iron can be made from ________ oxide by heating it with ________. The carbon ________ the iron oxide, forming carbon ________.

2 **Reorder these sentences to describe the three-stage reaction that takes place in the blast furnace.**

- **Carbon dioxide and carbon react, giving carbon monoxide.**
- **Carbon monoxide reduces the iron oxide.**
- **Coke burns in a blast of air to make carbon dioxide.**

3 **Why is limestone added to the mixture in the blast furnace?**

Summary

- Iron oxide is reduced by carbon to iron.
- A very high temperature is needed before this reaction occurs. This is reached in a blast furnace.
- The reaction in a blast furnace has three stages.

5:7 Electrolysis

What about the reactive metals?

Carbon reduction is a relatively cheap and easy way to get the less reactive metals from their ores.

Unfortunately, carbon reduction does not work on the more reactive metals such as aluminium, which is above it in the reactivity series.

These metals have to be 'torn' from their compounds using electrical energy – lots of it. This is a very expensive option and is only used when there is no other choice.

a **List four metals other than aluminium that cannot be made by carbon reduction.**

b **Iron and copper can be torn from their ores using electrical energy too. Why is this method not used in industry?**

very reactive	**potassium**
	sodium
	calcium
	magnesium
	aluminium
	carbon
	zinc
	iron
	tin
	lead
	hydrogen
	copper
unreactive	gold

How does this method work?

When metals react with non-metals the metal forms a positive ion. Common salt (sodium chloride, NaCl) contains positive sodium ions.

In solid salt, the ions are trapped. But if the salt is melted or dissolved in water the ions become free. They can be made to move.

the positive sodium (Na) ions are trapped in the solid salt

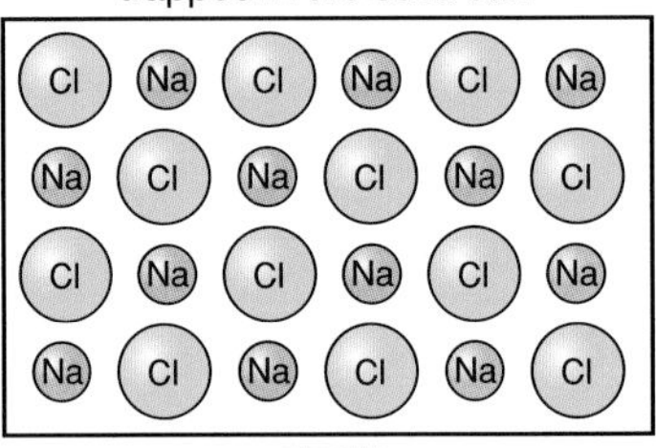

dissolve or melt

the sodium ions are free to move

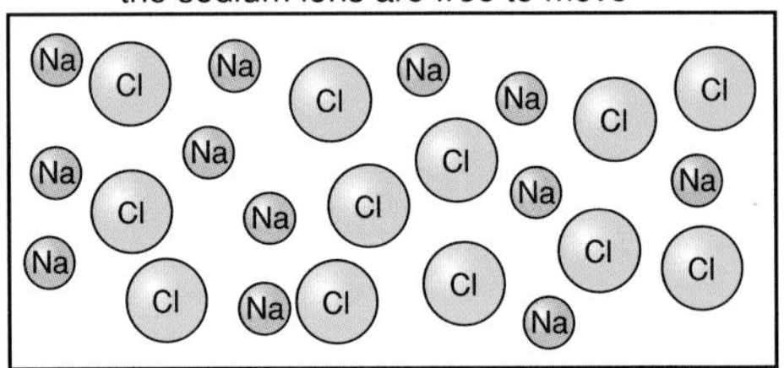

Sodium from molten salt

If you put electrodes into molten salt, the *positive* sodium ions are attracted to the *negative* electrode.

c **Copy and complete this phrase:**
Opposite charges __________.

When the positive sodium ions reach the negative electrode, the charges cancel out and the sodium ions turn back to sodium atoms. Metallic sodium forms on the electrode. This process is called **electrolysis**.

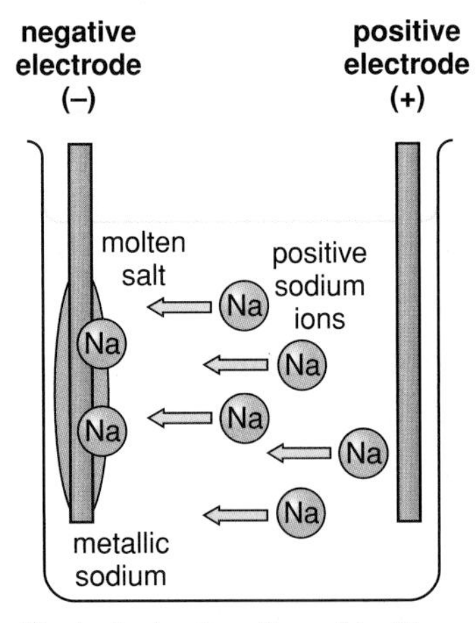

Electrolysis of sodium chloride.

Copper from copper sulphate solution

Electrolysis can work in solution, too.

Copper has positive ions in copper sulphate. These ions give copper sulphate its blue colour. In copper sulphate solution they are free to move.

If you put electrodes into copper sulphate, the *positive* copper ions are attracted to the *negative* electrode. When they reach the electrode, the charges cancel out and the copper ions turn back to copper atoms. Pure metallic copper is deposited on the electrode.

Copper is made from its ore by carbon reduction in a blast furnace. But the copper produced is not pure enough to use for electric wires. Electrolysis is used to purify the copper from the blast furnace.

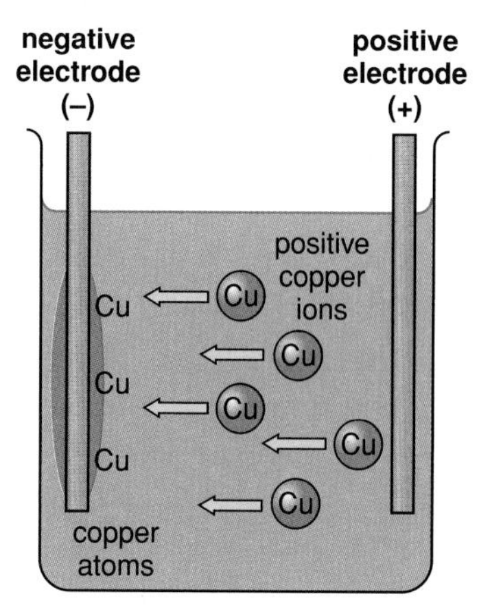

Electrolysis of copper sulphate solution.

d Why are the copper ions attracted to the negative electrode?

e What would happen to the colour of the solution if all the copper ions turned back to metal atoms at the electrode?

Electroplating

Gold is a very expensive metal – that is why gold jewellery costs so much.

A cheaper method is to make the jewellery from a less expensive metal, such as copper. The copper can then be coated with a thin layer of gold to make it look nice.

This coating is done by putting the copper jewellery into a solution containing positive gold ions. The copper jewellery is connected up as the negative electrode. The gold ions are pulled towards it and turn back into gold atoms. Pure gold is plated onto the surface of the copper.

Questions

1 Copy and complete the following sentences. Choose from:

positive **electrode** **electrolysis**
molten **Reactive**

__________ metals are torn from their compounds by __________. This works because metals form __________ ions, which are attracted to the negative __________. The metal compound has to be __________ so that the ions are free to move.

2 Potassium can be made by electrolysing molten potassium chloride.

a Why does it have to be molten?

b Are potassium ions positive or negative?

c Which electrode will the potassium form around?

Summary

- Reactive metals such as aluminium have to be torn from their ores using electricity. This process is called electrolysis.
- The metal collects at the negative electrode.
- Electrolysis is expensive, so it is only used when other methods cannot work.
- Copper is purified by electrolysis using copper sulphate solution.

5:8 Industrial electrolysis

Aluminium is a very useful metal. You can use aluminium to build things which:

- are strong;
- yet are very light.

Just the metal for building aeroplanes.

a Why do the properties of aluminium make it suitable for aeroplane bodies?

Aluminium is the commonest of all the metals in the Earth's crust, so you might expect it to be cheap to make. But, as aluminium is more reactive than carbon, it cannot be made by carbon reduction. Aluminium has to be extracted by electrolysis, this makes it more expensive to make.

b Why might you have expected aluminium to be cheaper than iron? Why is this not the case?

Making aluminium

Aluminium is made by the electrolysis of aluminium oxide. Aluminium oxide is purified from its common ore, **bauxite**.

Aluminium oxide is an ionic compound. Before it can be electrolysed, it must be melted. Unfortunately, aluminium oxide has a very high melting point.

Electrolysis is already an expensive process. Having to reach such high temperatures would make it even more expensive.

Fortunately another aluminium compound, called **cryolite**, melts at a lower temperature. The bauxite is dissolved in a small amount of molten cryolite. This lowers the melting temperature to just 950 °C. Even so, this still takes vast amounts of energy.

c Why does the aluminium oxide have to be molten?

d How does cryolite help to keep down the cost of your drink cans?

How it works

Electrolysis of aluminium takes place in specially designed units. The raw materials are fed into the unit, where they melt.

Positive aluminium ions are attracted to the negative electrode – which in this case is the container itself. Here they turn into aluminium atoms. The molten aluminium is run off at the bottom.

The oxygen in aluminium oxide forms negative ions. These negative oxygen ions are attracted to the positive electrode (carbon blocks). There they turn back to oxygen gas.

The carbon blocks burn away in the oxygen to make carbon dioxide gas. They have to be replaced regularly.

e Why do the carbon blocks of the positive electrode have to be replaced regularly?

Electrolysis of aluminium.

Because of the high production costs involved, aluminium plants tend to be very large. It is cheaper (per tonne) to make a lot rather than a little. A typical plant may have 300 electrolysis cells like the one shown, each making a tonne of aluminium every day.

Aluminium plants often need a power station of their own, as they use as much electricity as a small town. No wonder aluminium costs more than iron.

f Why can't aluminium plants just plug into the mains like everyone else?

Questions

1 Copy and complete the following sentences. Choose from:

positive negative electrolysis molten atoms

Aluminium is produced from __________ aluminium oxide by __________. The positive aluminium ions move to the __________ electrode and turn back to aluminium __________. The negative oxygen ions go to the __________ electrode and turn to oxygen gas.

2 Aluminium is too reactive to electrolyse in solution. How is it that copper can be purified from solution?

Summary

- Aluminium is made by the electrolysis of molten bauxite.
- Cryolite is added to lower the melting point.
- This is still an expensive process.

5:9 Corrosion

Iron and steel are very useful metals, but they have one major fault: they rust!

Why does iron rust?

Iron and steel corrode by reacting with air (oxygen) and water.

$$\text{iron} + \text{oxygen} \xrightarrow{\text{water}} \text{iron oxide (rust)}$$

The simplest way to stop iron or steel rusting is to paint it. The paint forms a barrier, stopping water and oxygen from getting at the iron. New cars have several layers of special paint to stop them rusting.

Oil, grease or a thin coat of plastic may also be used as a barrier to prevent rust.

a Why must scratches or chips on car paintwork be painted over straight away?

b Steel wire garden fencing is often coated in green plastic to make it look nice. In what other way does this plastic help?

Tin cans and barbed wire

Sometimes steel is coated with another metal to make a barrier and protect it. Steel food cans are coated with tin. Tin is less reactive than iron. If the tin gets scratched, the iron soon rusts away.

Barbed wire and some nails are coated with zinc. If the zinc layer is scratched, the iron still does not rust. This is because zinc is more reactive than iron. This makes the zinc corrode *instead* of the iron.

c Why does zinc-plated steel stay rust-free longer than tin-plated steel?

It's the iron that rusts in a rusty 'tin' can.

What a sacrifice!

Iron and steel rust quickly in salt water, so steel-plated ships need a lot of protection. The idea that a more reactive metal will corrode instead of steel can be used to provide this protection.

Zinc or magnesium blocks are bolted onto the hull of the ship. Over the months at sea these steadily corrode away, leaving the steel rust-free. When nearly all the zinc has gone, a new block is simply bolted on in its place. This is called **sacrificial protection** because the zinc is sacrificed to save the iron. Sacrificial protection is a cheap and easy way to protect the hull.

When this zinc block has corroded it will need replacing – but the hull will be undamaged by rust.

Stainless steel

Steel is an **alloy** of iron and carbon. If iron is mixed with other metals, other useful alloys are formed.

If chromium is mixed with iron, an alloy forms that does not rust at all. This is called stainless steel. It is used for cutlery – and some expensive cars.

d Can you think of any reasons why all cars aren't made from stainless steel?

The world's first stainless steel roof on the Chrysler Building, New York.

Stainless aluminium?

Aluminium is high in the reactivity series. Because of this, you might expect it to corrode very quickly. Yet aluminium foil stays bright and shiny. Why is that?

Aluminium is so reactive that it quickly forms a thin, invisible oxide layer on its surface. The oxide layer sticks so firmly to the metal that it acts as a barrier and prevents further corrosion.

Aluminium also forms alloys. It is mixed with magnesium to make it strong enough to build aeroplanes.

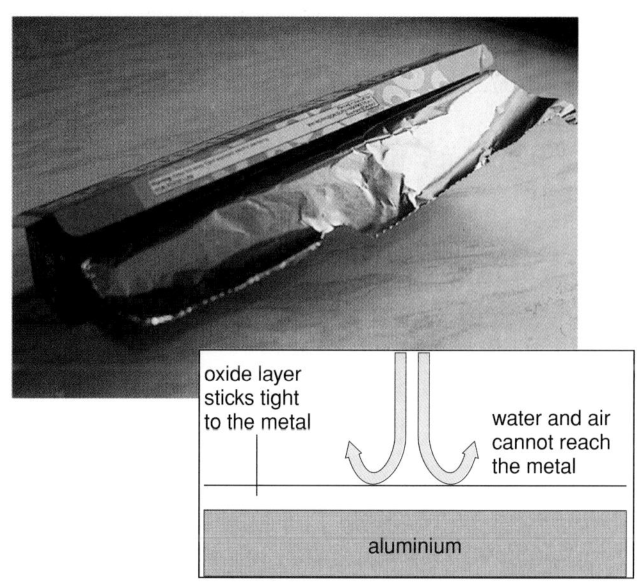

Aluminium stays shiny because of its protective oxide layer.

Questions

1 Copy and complete the following sentences. Choose from:

corrode **reactive** **zinc** **sacrificial**

If a block of __________ is bolted onto steel, the zinc will __________ instead of the steel. This is called __________ protection. It works because zinc is more __________ than steel.

2 Which of these metals could be bolted onto a ship to protect its steel hull?

- Lead.
- Tin.
- Aluminium.

3 Copper is less reactive than iron. What do you think would happen if a block of copper was bolted to the hull instead of zinc?

4 Knives, forks and spoons are made of an alloy of iron. Which other metal is used in this alloy?

5 How does aluminium oxide protect aluminium from corrosion?

Summary

- Iron rusts in the presence of oxygen and water.
- Paint acts as a barrier and so prevents rusting.
- Zinc blocks stop rust by sacrificial protection.
- Aluminium is protected from corrosion by a thin oxide layer.

5:10 Acids and alkalis

Alkali metal hydroxides dissolve in water. The solution they make is strongly alkaline. Alkalis are the opposites of **acids**.

The *strength* of an acid or alkali is measured against the **pH scale**. This scale runs from 1 to 14.

- 1 is the strongest acid.
- 14 is the strongest alkali.
- Water has a pH of 7. This is neither acid nor alkali – it is **neutral**.

strongest acid | neutral | strongest alkali

1	2	3	4	5	6	7	8	9	10	11	12	13	14

hydrochloric acid | vinegar | acid rain | pure water | soap | ammonia solution | sodium hydroxide

pH value as shown by universal indicator.

The pH of a solution may be found using **universal indicator**. This changes through many different colours from pH 1 to pH 14.

a Name a weak acid.

b Is soap acid or alkaline? How strong is it?

Making salts

Acids and alkalis are chemical opposites. Mixed together in the right proportions they can cancel one another out. This process is called **neutralisation**.

Two new compounds are formed in a neutralisation reaction: water, and a type of compound called a **salt**.

acid + alkali → salt + water (neutral)

The name of this salt depends on:

- the metal in the alkali;
- the acid used.

With sodium hydroxide and hydrochloric acid, the salt is sodium chloride – *common salt*.

sodium hydroxide + hydrochloric acid → sodium chloride + water

c What is happening to the acid in the photographs? How do you know?

d What is the name of the salt that forms?

Did you know?
Your stomach contains hydrochloric acid to help digest your food! If this gets too strong you get acid indigestion. This can be cured by chewing 'antacid' tablets, made from magnesium hydroxide.

If you drop an antacid tablet into a glass of hydrochloric acid containing universal indicator, the colour changes. This shows that the acid is becoming weaker, until there is none left.

Name that salt ...

- If you neutralise hydrochloric acid you get *chlorides*.
- If you neutralise sulphuric acid you get *sulphates*.
- If you neutralise nitric acid you get *nitrates*.

So, if you neutralise potassium hydroxide with sulphuric acid you get potassium sulphate:

potassium hydroxide + sulphuric acid → potassium sulphate + water

If you neutralise lithium hydroxide with nitric acid you get lithium nitrate:

lithium hydroxide + nitric acid → lithium nitrate + water

e **Potassium nitrate is used to make gunpowder. How could you make this salt? Write a word equation for this.**

The strange case of ammonia ...

Ammonia is a gas with a very strong smell. It is a simple compound of the non-metals nitrogen and hydrogen. The formula of ammonia is NH_3. When it dissolves in water, it forms a compound called ammonium hydroxide:

ammonia + water → ammonium hydroxide

Ammonium hydroxide solution is a reasonably strong alkali. It is often used as a kitchen cleaner, as alkalis are good for clearing up grease. What is odd about ammonia is that the 'ammonium' (NH_4) part is acting as if it were an alkali metal. Ammonium hydroxide can be used to neutralise acids, forming 'ammonium' salts with them:

ammonium hydroxide + nitric acid → ammonium nitrate + water

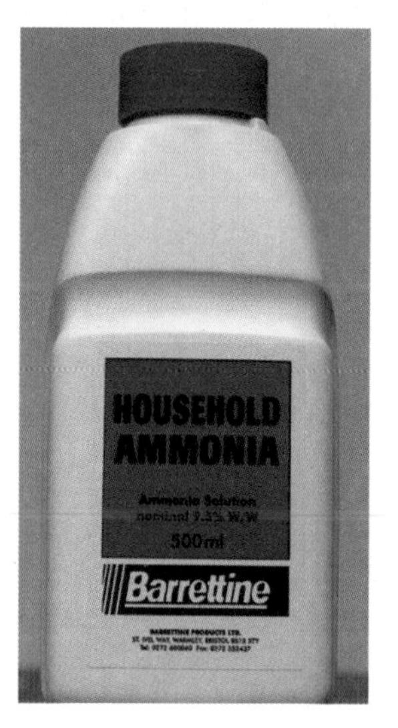

f **Salts are usually made from metals and acids. What is odd about ammonium nitrate?**

Questions

1 **Copy and complete the following sentences. Choose from:**

alkali indicator green red

Universal __________ turns __________ in acid. If an __________ is added drop by drop the colour gradually changes. When the alkali has neutralised the acid, the colour will be __________.

2 **Copy and complete the following equation:**

acid + alkali → __________ + water

3 **What salt will you get if you neutralise ammonia solution with sulphuric acid? Write a word equation for this reaction.**

Summary

- The strength of acids and alkalis can be measured against the pH scale.
- Universal indicator changes colour with pH.
- Salts are made when an acid is neutralised by an alkali such as sodium hydroxide.
- Ammonia also forms an alkaline solution and can make salts.

5:11 More salts

Transition metal hydroxides

Transition metals form hydroxides, but these are not soluble in water. They do *not* therefore form alkalis.

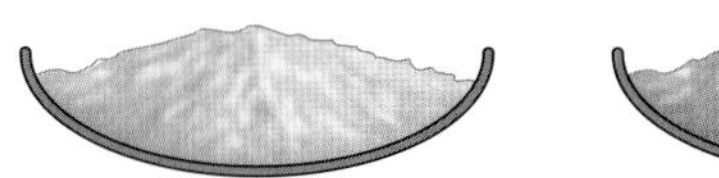
copper hydroxide

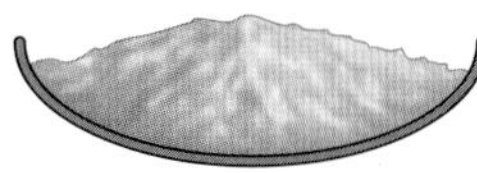
iron hydroxide

a Why are the two hydroxides in the diagram not alkalis?

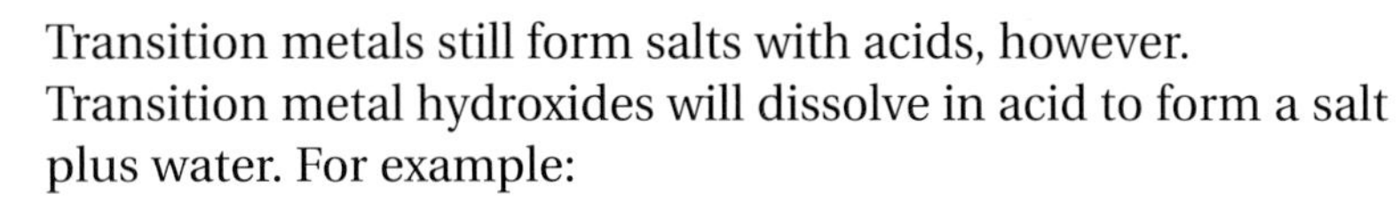

Transition metals still form salts with acids, however. Transition metal hydroxides will dissolve in acid to form a salt plus water. For example:

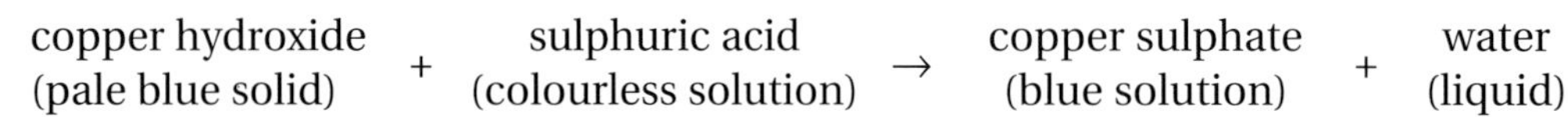

copper hydroxide (pale blue solid)	+	sulphuric acid (colourless solution)	→	copper sulphate (blue solution)	+	water (liquid)

b What salt would you get if you dissolved copper hydroxide in hydrochloric acid?

> **Did you know?**
> A solution in water is called an **aqueous solution.**

Oxides too

Transition metal oxides will also dissolve in acid like this:

copper oxide (black solid)	+	sulphuric acid (colourless solution)	→	copper sulphate (blue solution)	+	water (liquid)

c What salt would you get if you dissolved copper oxide in hydrochloric acid?

Bases

Oxides and hydroxides that react with acids are called **bases**. Alkalis are simply bases that dissolve in water.

The reaction between an acid and a base is:

acid + base → salt + water

d Which of these compounds are bases?

- potassium hydroxide
- iron oxide
- chromium hydroxide

What's going on?

Acids contain hydrogen (H). Bases contain oxygen (O). (Hydroxides (OH) have hydrogen as well as oxygen.)

When acids and bases react, the hydrogen and oxygen form water (H_2O). The salt is made from the bits left over!

e How is water made when an acid and a base react?

Hydrogen from the acid.

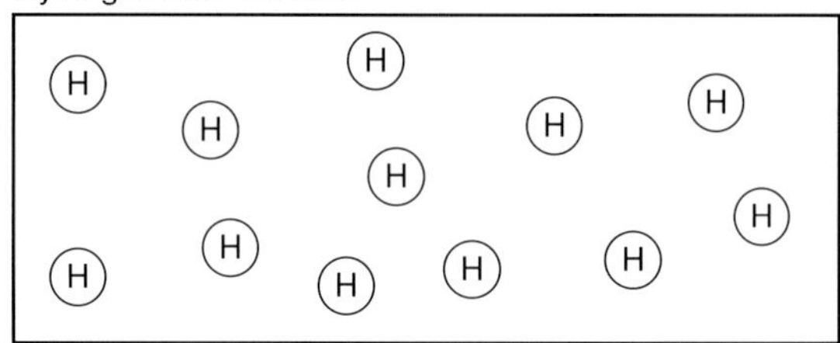

Oxygen from the base.

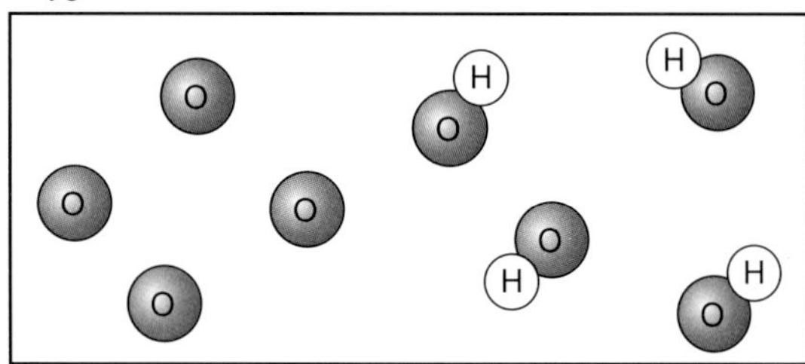

Hydrogen and oxygen form water.

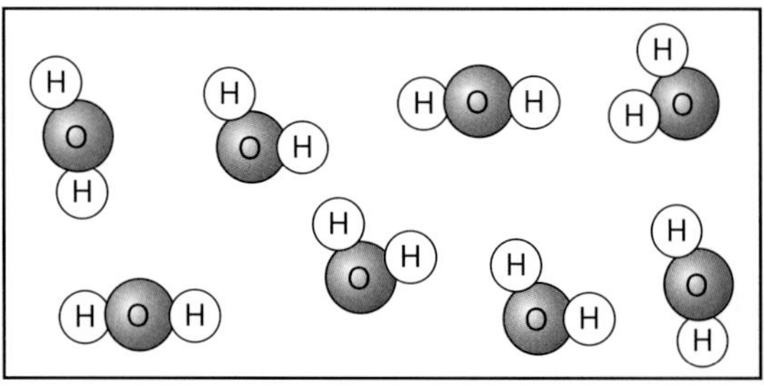

Making water.

Making copper sulphate

Transition metal salts can be made in this way. For copper sulphate:

stage 1

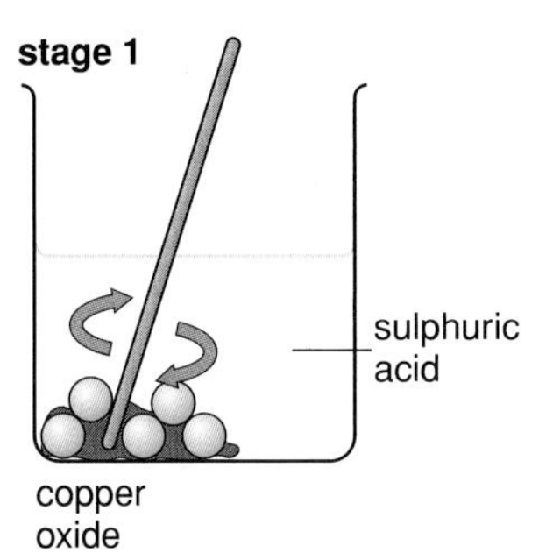

Add copper oxide (or copper hydroxide) to sulphuric acid and stir until it has all dissolved.

stage 2

copper sulphate solution

undissolved copper oxide

Add more copper oxide a little at a time until no more will react and dissolve.

stage 3

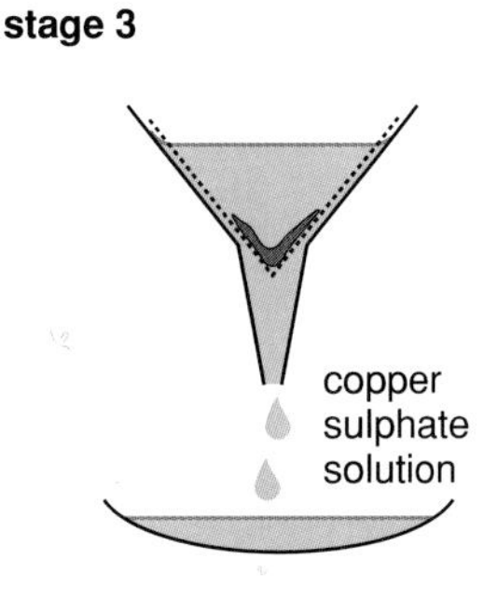

Filter out the unused copper oxide. The clear filtrate will be copper sulphate solution.

stage 4

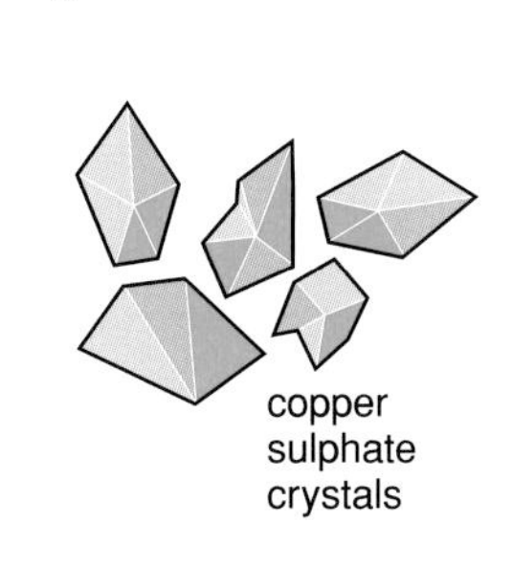

Let the water evaporate away to give blue copper sulphate crystals.

Making copper sulphate.

f How do you know when the acid has been neutralised?

g Describe how you could make nickel sulphate from insoluble nickel oxide.

Useful salts

Transition metal salts have many uses.

- Copper sulphate is sprayed onto grape vines to prevent disease.
- Copper chloride is used to cure fish fungus.
- Iron sulphate is used in iron tablets. People take these to keep their blood in good condition.
- Silver salts are used in black and white photography.

Copper sulphate in use at a vineyard.

Questions

1 Copy and complete the following sentences. Choose from:

alkalis bases neutralise oxides hydroxides

All metal __________ and hydroxides are __________. These can __________ an acid to form a salt and water. Only __________ that dissolve in water are __________.

2 Copy and complete this equation:

chromium oxide + hydrochloric acid →

__________ + __________

3 Copy and complete this equation:

copper oxide + __________ acid →

copper nitrate + __________

Summary

- Transition metal hydroxides are not soluble in water so they are not alkalis.
- Metal oxides and metal hydroxides are bases.
- acid + base → salt + water

End of module questions

1 This diagram shows the periodic table.

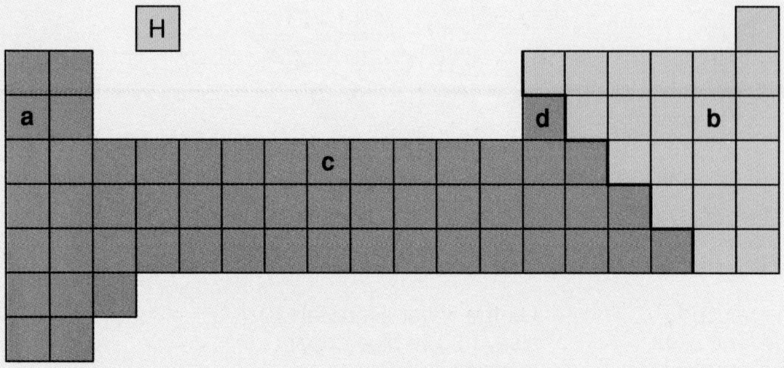

1.1 Which letter shows the position of a non-metal?

1.2 Which letter shows the position of a transition metal?

1.3 Which letter shows the position of a metal in group 3?

1.4 Which letter shows the position of an alkali metal?

2 The sentences below are about the alkali metals. Use the words below to complete the sentences.

ionic

density

reactive

electricity

2.1 Alkali metals are good conductors of __________.

2.2 Sodium floats on water because it has a low __________.

2.3 Potassium is a very __________ metal.

2.4 Alkali metals react with non-metals to form __________ compounds.

3 The sentences below are about the transition metals such as iron or copper. Choose the correct word from each pair given.

3.1 Transition metals are (good/bad) conductors of electricity.

3.2 Transition metals are usually much (stronger/weaker) than the alkali metals.

3.3 Transition metal salts are usually (white/coloured).

3.4 Transition metals are (less/more) reactive than alkali metals.

4 Here are four metal compounds.

A copper sulphate
B sodium hydroxide
C copper oxide
D sodium chloride

4.1 Which of these compounds dissolves in water to give an alkaline solution?

4.2 Which of these compounds is a base but not an alkali?

4.3 Which of these compounds is a blue-coloured salt?

4.4 Which of these compounds can be made using hydrochloric acid?

5 The sentences below are about how metals are produced. Choose the correct word or phrase from each pair given.

5.1 Iron can be made from iron oxide in a blast furnace using coal because it is (more/less) reactive than carbon.

5.2 Gold was one of the first metals to be discovered because it is so (reactive/unreactive).

5.3 Aluminium has to be made by electrolysis because it is (more/less) reactive than carbon.

5.4 Copper is *purified* by (electrolysis/carbon reduction).

6 Choose the two best answers.

6.1 Look at this reactivity series:

magnesium	most reactive ↑
zinc	
iron	
lead	
hydrogen	
copper	
silver	least reactive

A Copper can displace silver from silver compounds.
B Lead corrodes faster than iron.
C Magnesium was one of the first metals to be discovered.
D Zinc will fizz in acid.
E Copper will fizz in acid.

6.2 Potassium hydroxide may be neutralised by nitric acid.

A Universal indicator will turn purple when the potassium hydroxide has been neutralised.
B Universal indicator will turn green when the potassium hydroxide has been neutralised.
C The formula for potassium hydroxide is CuO.
D The salt that forms in this reaction is potassium sulphate.
E The salt that forms in this reaction is potassium nitrate.

7 Iron is produced in a blast furnace. The raw materials are coke, iron ore and limestone.

7.1 When iron oxide is converted to metallic iron in this way the iron is ...

A oxidised.
B reduced.
C electrolysed.
D neutralised.

7.2 A blast of hot air is used to ...

A blow out the waste gases.
B oxidise the iron.
C help to form the slag.
D make the coke burn fiercely.

7.3 When the carbon dioxide reacts with more carbon it forms ...

A sulphur dioxide.
B metallic iron.
C coke.
D carbon monoxide.

7.4 Limestone is added to ...

A speed up the reaction.
B remove the acid impurities.
C make the iron melt.
D absorb the carbon dioxide.

8 Aluminium is produced by the electrolysis of molten bauxite (aluminium oxide).

8.1 When bauxite is converted to aluminium in this way the aluminium is ...

A displaced.
B oxidised.
C reduced.
D neutralised.

8.2 In molten bauxite, aluminium is in the form of ...

A freely moving positive aluminium ions.
B freely moving negative aluminium ions.
C aluminium oxide molecules.
D trapped positive aluminium ions.

8.3 Cryolite is added to the bauxite because ...

A it is cheaper than bauxite.
B it reacts with the oxygen ions.
C it lowers the melting point.
D it reduces the aluminium.

8.4 The carbon blocks used as the positive electrodes have to be replaced regularly because ...

A they melt in the molten bauxite.
B they do not conduct electricity.
C they burn off in the oxygen that is formed there.
D they form negative carbon ions.

9 These questions are about the way metals corrode and how corrosion may be prevented.

9.1 Simply painting over rust is not a solution to the problem because ...

A paint keeps the water out but lets the oxygen through.
B paint keeps the oxygen out but lets the water through.
C rust reacts with the paint and dissolves it.
D rust swells and cracks the paint, letting oxygen and water back in.

9.2 Iron corrodes much more rapidly than aluminium because ...

A aluminium is more reactive than iron.
B aluminium is less reactive than iron.
C aluminium oxide forms a protective layer.
D aluminium oxide simply rubs off.

9.3 A zinc block can be used to prevent the steel hull of a ship from rusting because ...

A the zinc stops the water getting to the steel.
B zinc is more reactive than steel and so reacts instead of it.
C zinc is less reactive than steel and so protects it.
D zinc is an inert metal.

9.4 Stainless steel is made by ...

A melting iron and carbon together.
B melting iron and chromium together.
C coating iron with chromium.
D coating iron with tin.

Module 6 – Earth materials

The earth is our spaceship, and it provides everything we need. The rocks of the Earth provide the raw materials we need for building or manufacturing goods as well as the soil in which we grow our food, while the atmosphere provides the oxygen we need to breathe. Yet the Earth can be a dangerous place, with earthquakes and volcanoes causing destruction in many parts of the world.

You already know about the main types of rock and how they are formed. In this module you will learn more about the raw materials we get from the Earth and how they are used. You will also learn more about the way the Earth works and how it has changed over its $4\frac{1}{2}$ billion year history. You will find out more about:

- crude oil: how it forms and what we can make from it;
- how the Earth's atmosphere has come to be as it is;
- how the theory of plate tectonics helps to explain all the mountain ranges, earthquakes and volcanoes;
- how scientists slowly pieced together this theory from the evidence in the rocks.

Before you start, check what you remember about rocks and how they form.

1 What type of rock forms when molten rock (magma) cools and sets?
2 Why are the crystals in lava so tiny?
3 What happens to rocks when they are weathered and eroded?
4 How are sedimentary rocks formed?
5 How are sedimentary rocks turned into metamorphic rocks?
6 What is the rock cycle?

6:1 Limestone

Limestone is a common rock with many uses. It usually formed in the sea from the remains of shells.

In Britain alone, 150 million tonnes of limestone are quarried every year. This is used:

- for buildings or road 'chippings';
- to neutralise excess acid in lakes or soils;
- to make cement and concrete.

a Limestone is very useful – but would you like to live near a large quarry like this? List three possible problems.

You will find huge limestone quarries like this in many parts of the country.

Using limestone

Limestone has been used as a building stone for thousands of years.

Today, most limestone is used as a **raw material** to make something new. It is changed by chemical processes into other useful products, such as cement.

Processing limestone

Limestone is the compound calcium carbonate. Each particle has one calcium atom, one carbon atom and three oxygen atoms: $CaCO_3$. All carbonates have one carbon atom and three oxygen atoms like this.

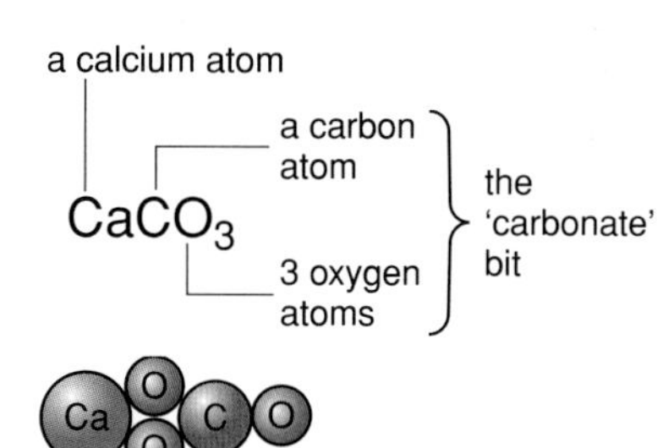

A model of a calcium carbonate particle.

If limestone is heated strongly, the compound breaks up and the atoms are rearranged:

- the carbon atom takes two oxygen atoms to form carbon dioxide – a gas;
- the calcium atom is left with just one oxygen atom – this is calcium oxide, which is also called **quicklime**.

$$\text{calcium carbonate} \xrightarrow{\text{heat}} \text{calcium oxide} + \text{carbon dioxide}$$

Breaking a compound by heating it like this is called **thermal decomposition**.

b Where does the carbon dioxide go?

c The solid calcium oxide left behind weighs less than the calcium carbonate. Why?

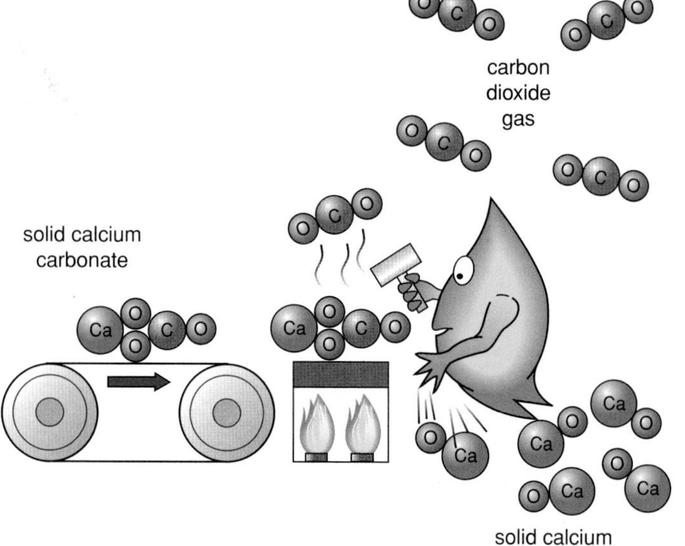

Thermal decomposition.

A simple fertiliser

Quicklime is a very strong alkali. It reacts with water, forming calcium hydroxide (**slaked lime**):

calcium oxide + water → calcium hydroxide + heat energy

Farmers have used slaked lime for centuries as a simple fertiliser to make their soil less acid. It also helps to break up the soil so that plants can grow well.

Sometimes lakes are polluted by acid rain. Slaked lime is used to get rid of the acid there as well.

> **Did you know?**
> The murderer Dr Crippen got rid of his victim by dissolving parts of her body in quicklime. It is such a strong alkali that it dissolved all but a few pieces of bone, teeth and skin.

Cement and concrete

Today, most limestone is used to make **cement**. The limestone is heated with clay in a big oven that turns to mix them up – a **rotary kiln**.

The mixture is then ground to form a light grey powder, which is the cement. Cement turns back to a hard solid when you add water. This is a slow chemical reaction.

Most cement is mixed with sand and gravel to make **concrete**. This is cheaper and stronger than pure cement.

Concrete forms a thick liquid when first mixed with water, and can be poured into any shape. Slow chemical reactions make it set after a few hours and eventually it becomes rock hard.

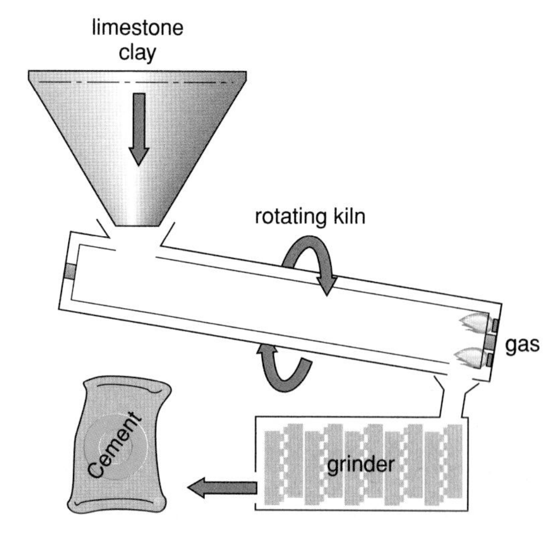

A rotary kiln.

d What are the raw materials for cement?

e Why is it easier to use concrete for building than solid limestone?

Making glass

Other useful materials are made using limestone. Glass is made by heating very pure sand (silicon dioxide) with limestone and a little soda (sodium carbonate). The chemicals react, melt, and carbon dioxide is given off. When this liquid cools, it forms a hard, but brittle, transparent material – glass!

Questions

1 Make a list of five different uses of limestone.

2 Copy and complete the following sentences. Choose from:

thermal **oxide** **carbon** **carbonate**

Limestone is made from calcium __________. When it is heated, this compound is broken up into calcium __________ and __________ dioxide gas. This is called __________ decomposition.

3 What happens when water is added to quicklime?

4 Describe simply how cement and concrete are made.

5 What raw materials are used to make glass?

Summary

- Limestone is used for building.
- Powdered limestone can neutralise acid.
- Limestone is used as a raw material to make cement.
- Carbonates can be broken up by heat. This is called thermal decomposition.

6:2 Oil

There she blows!

Millions of gallons of **crude oil** gush from a hole in the ground. Time to celebrate if you have just drilled a well and made a lucky strike. This black liquid will make you rich.

It can be used as a raw material to make:

- petrol and other fuels;
- plastics;
- paints and solvents;
- medicines.

a But what about the environment! List three ways in which crude oil causes environmental problems.

How did it form?

Crude oil is a **fossil fuel**. Oil and natural gas are formed when plant and animal remains are buried deep underground. Over millions of years, heat and pressure change them into oil and gas.

Coal is another fossil fuel, which forms when trees are buried.

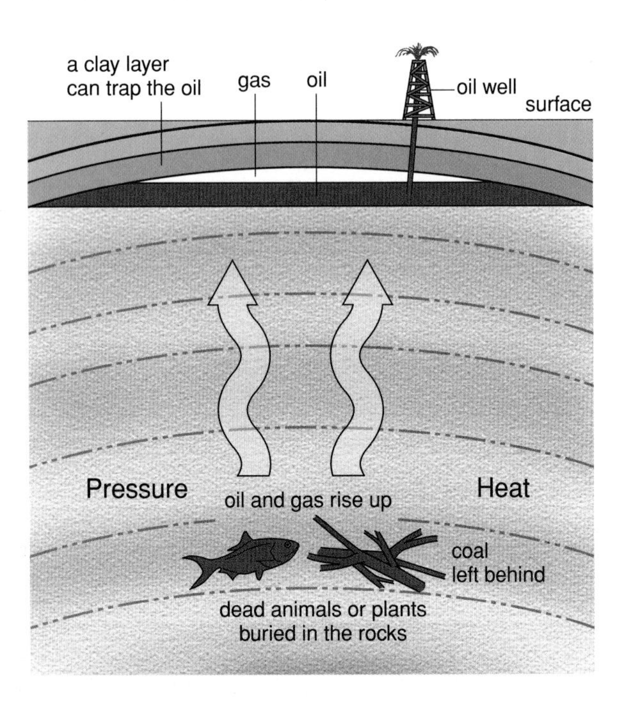

b A fuel is something we can get energy from. A fossil is the remains of a plant or animal found in the rock. Why are coal, oil and gas called fossil fuels?

c What would happen to the oil and gas if there was not a clay layer to trap it underground?

What is crude oil made of?

Crude oil is a mixture of many different compounds. As they are not chemically combined, each compound has different physical properties.

On their own, some compounds are:

- runny liquids;
- thick liquids;
- solids;
- gases.

Mixed up together they are a thick, gooey black liquid.

As a mixture, crude oil is useless. The different compounds must be separated out before they can be used.

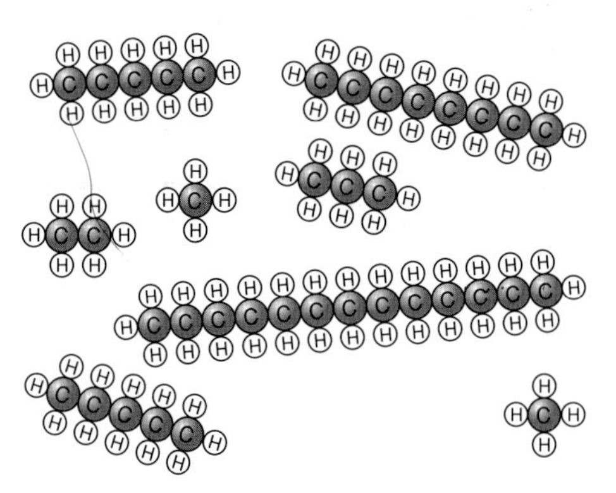

Crude oil is a mixture.

d List five other mixtures you know from home or school.

What kind of compounds?

Most of the compounds found in crude oil are made from two types of atom only: hydrogen and carbon. Compounds like this are called **hydrocarbons**.

These hydrocarbons have molecules of different sizes. Carbon atoms can form chains. Big molecules have long chains; small molecules have short chains.

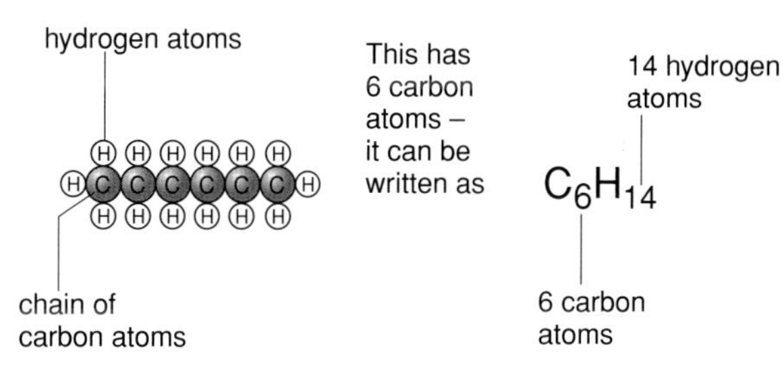

Hexane, a hydrocarbon.

e Look at the crude oil mixture at the top of the page. How many carbon atoms are there in the chain of the longest hydrocarbon?

Questions

1 Copy and complete the following sentences. Choose from:

plants heated millions

Crude oil formed from the remains of animals and ________ that lived ________ of years ago. They were buried in the rocks, where they were squashed and ________.

2 Why are the compounds in crude oil called hydrocarbons?

3 Draw a hydrocarbon molecule with three carbon atoms in its chain.

4 Crude oil is a mixture of different compounds of hydrogen and carbon. Explain the terms *mixture* and *compound*.

Summary

- Crude oil is a fossil fuel.
- It formed from plants and animals that were buried in the rocks millions of years ago.
- Crude oil is a mixture of hydrocarbons.

6:3 The chain gang

All in the family

Most of the compounds in crude oil belong to the hydrocarbon family. This means that they have many similar properties:

- most are liquids;
- they can be used as fuels because they burn in air.

a What gas are hydrocarbons reacting with when they burn in air?

In all families there are differences, too. In the hydrocarbon family, the differences are linked to the length of the carbon chain.

How runny?

When liquids flow, the molecules move over one another.

- Hydrocarbons with small molecules (short chains) can move over one another easily. They form very runny liquids, like water.
- Hydrocarbons with large molecules cannot move over one another so easily. The long carbon chains 'stick' against one another, so the liquid seems very 'thick', like syrup.

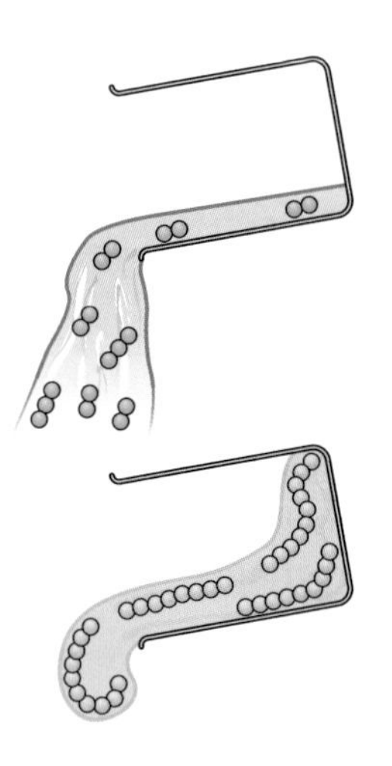

Liquids that are 'thick' and slow moving like this are called **viscous** liquids.

b Are the following liquids runny or viscous?

wine	treacle	milk
vinegar	syrup	

Boiling point

For a liquid to boil, the molecules have to move fast enough to break free from one another. The molecules move faster when the liquid is heated.

- Short-chain hydrocarbons boil at low temperatures.
- Long-chain hydrocarbons must be heated to much higher temperatures before they boil.

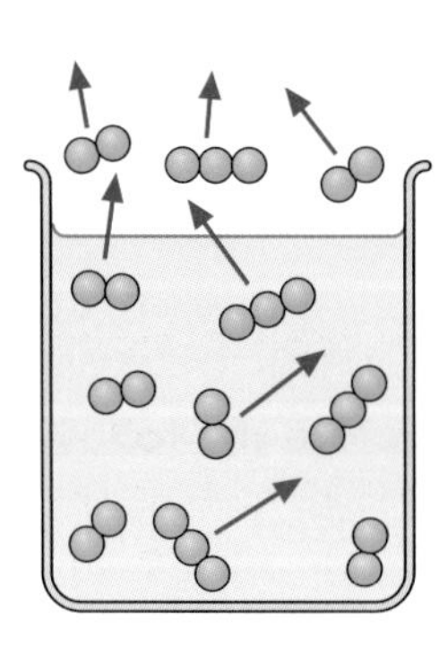

c A hydrocarbon with a very short chain of just four carbon atoms boils at 0 °C. Would this be a solid, a liquid or a gas at room temperature?

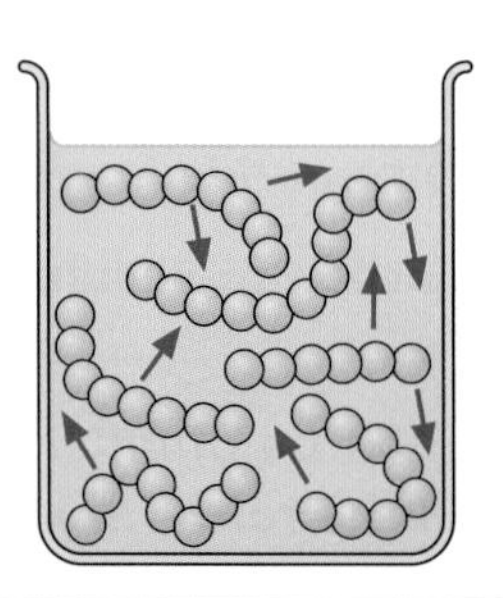

Can you smell it?

Some molecules can escape from a liquid below its boiling point. That's why you can smell perfume without having to boil it! Liquids that easily turn into a gas like this are called **volatile** liquids.

The lower the boiling point of a liquid, the more volatile it is.

d Petrol has short-chain hydrocarbons and so has a low boiling point. Will petrol be a volatile liquid or not?

Useful or dangerous?

Hydrocarbon fuels have to be turned into a gas before they are burnt. Petrol is a useful fuel because it is so volatile. It is easy to turn it into a gas. Less volatile hydrocarbons with longer chains are not so easy to use as a fuel.

But there is a downside to this. Spilt petrol will quickly turn into a dangerously explosive gas. A single spark or flame could set off an explosion. No wonder there are 'No Smoking' signs at petrol stations.

e In some countries with few petrol stations, small amounts of petrol are sold in old glass or plastic bottles from roadside stalls. Do you think this is safe? Explain your answer.

Questions

1 Copy and complete the following sentences. Choose from:

runny **boiling** **hydrocarbons** **viscous** **length**

The properties of __________ vary with the __________ of the carbon chain. Those with short chains are __________ liquids and have low __________ points. Those with longer chains are __________ liquids and have high boiling points.

2 Fuel oil is a slightly viscous liquid that is sometimes used for home heating. It can be safely poured into a tank without risk of explosion. Does it have a longer or shorter carbon chain length than petrol? Explain your answer.

3 Draw a bar chart to show how boiling point changes with hydrocarbon chain length using the data shown in the table.

Carbon chain length	5	6	7	8	9
Boiling point (°C)	36	69	99	125	151

What happens to the boiling point when the chain length gets longer?

Summary

As hydrocarbon molecules get bigger (have longer carbon chains):

- they become more viscous;
- their boiling points get higher;
- they get less volatile;
- they catch fire less easily.

6:4 Separating out

Fractional distillation

Crude oil is a mixture of many useful hydrocarbons that need to be separated out before they can be used. This is done by a form of **distillation**.

If you heat crude oil enough it will boil. The gas can then be cooled so that it **condenses** back to a liquid for collection.

- Long-chain hydrocarbons with high boiling points will condense out quickly.
- Short-chain hydrocarbons with low boiling points will have to be cooled down much more before they condense.

a Steam condenses back to water at 100 °C. Paraffin boils at 200 °C. At what temperature will paraffin gas condense back to a liquid?

As the gas cools, different **fractions** can be collected. The process is called **fractional distillation**.

The fractionating tower

In an oil refinery, crude oil is heated strongly in a furnace so that it *all* boils. The gas then passes up through a **fractionating tower**, which is hot at the bottom but cold at the top.

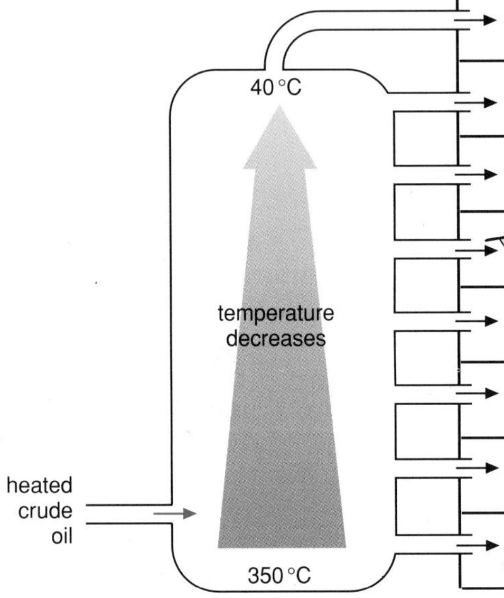

40 °C

temperature decreases

heated crude oil

350 °C

Fraction	Average number of carbon atoms
LPG (liquid petroleum gas) is used for Calor gas stoves.	3
Petrol is used as fuel for cars.	8
Naphtha is used to make chemicals.	10
Paraffin is used as a jet fuel as well as for paraffin stoves.	12
Diesel is used as a fuel for lorries.	20
Fuel oil is used for central heating.	40
Lubricating oil is used to lubricate machinery.	80
Bitumen is used to make roads.	120

The different hydrocarbons condense out at different levels of the tower and can be piped off.

b Where will short-chain hydrocarbons condense out – at the top or bottom of the tower?

c Bitumen collects at the bottom of the tower. Does bitumen have long or short chains?

The useful fractions

Each fraction is still a mixture. But the range of chain lengths within each fraction is small, so the hydrocarbons have very similar properties.

Petrol contains a range of molecules that have between 6 and 10 carbon atoms. The mixture as a whole is still volatile enough to work in a car engine.

d In which fraction would you find a 35-carbon hydrocarbon? What property makes this safer to store than petrol?

e LPG has to be stored under pressure in strong metal cylinders. Why would an LPG cylinder with a leaking valve be dangerous?

How much of each fraction?

From 100 000 tonnes of crude oil you might get:

- 25 000 tonnes of petrol;
- 15 000 tonnes of paraffin;
- 25 000 tonnes of diesel;
- 20 000 tonnes of fuel oil.

f Draw a pie chart to show the percentage of petrol, paraffin, diesel, fuel oil and 'other products' that are produced.

g Petrol makes up about 40% of all the crude oil products *sold* (by tonnage). Why might this cause problems for the oil refinery?

Question

1 Rearrange the sentences below to describe how fractional distillation works. Copy out this description.

- Long-chain hydrocarbons condense out at the bottom of the tower.
- The tower is hot at the bottom and cold at the top.
- Crude oil is turned into a gas in a furnace.
- Short-chain hydrocarbons condense out higher in the tower.
- The gas flows up through a fractionating tower.

Summary

- Crude oil is separated into useful products by fractional distillation.
- Each fraction has hydrocarbons of similar chain lengths and properties.

6:5 Cracking

The fractional distillation of crude oil makes many useful products. Unfortunately, the demand for these does not match the supply from the refinery.

The biggest demand is for short-chain hydrocarbons like those in petrol, so oil companies have plenty of long-chain hydrocarbons left over.

a There's not enough petrol from the refinery – but which fractions are there too much of?

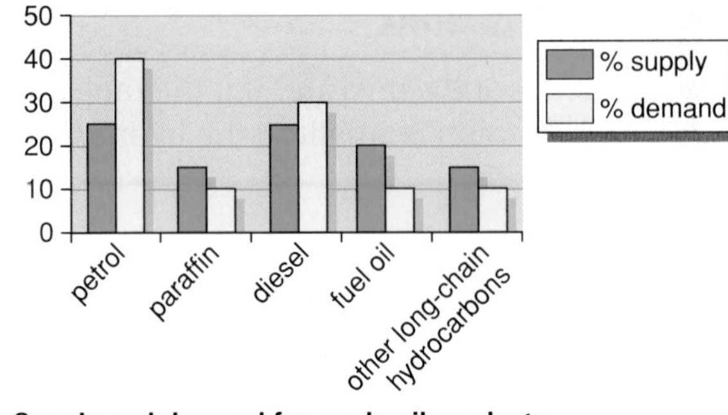

Supply and demand for crude oil products.

Cracking

If compounds are heated enough they can be broken up. This is called thermal decomposition.

Long-chain hydrocarbons can be broken into smaller, more useful pieces by thermal decomposition. When long-chain hydrocarbons are 'chopped up' like this it is called **cracking**.

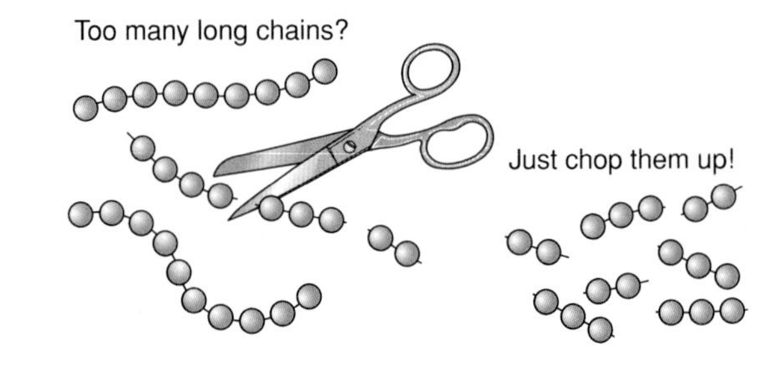

b If a 20-carbon hydrocarbon was cracked to make an 8-carbon hydrocarbon for petrol, how many carbon atoms would be left in the remaining molecule?

Industrial cracking

The cracking process is built into the oil refinery:

- the unwanted long-chain hydrocarbons are boiled;
- the vapour is passed over a hot **catalyst**;
- thermal decomposition takes place, making new short-chain hydrocarbons.

A catalyst is a chemical which speeds up a reaction. A catalyst is used here to make the thermal decomposition reaction go faster.

The short-chain products are passed back through the fractionating tower to separate them out.

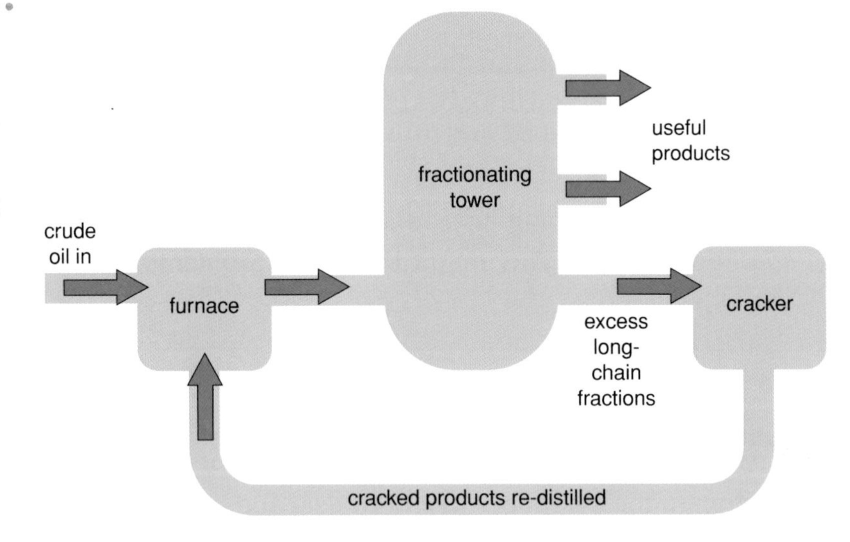

The extra petrol needed is produced in this way, as well as other important chemicals used to make plastics.

c Why do the products from the cracker have to be re-distilled?

WeDistill Oil Company

You are the production manager for the WeDistill Oil Company. Your job is to make sure you have the right amount of each product type to satisfy your customers.

Not all crude oil is the same. The amount of each fraction you get depends on where the oil comes from. You have a long-term contract to buy Arabian oil.

	Petrol (%)	Paraffin (%)	Diesel (%)	Long-chain (%)
Arabian crude oil	20	10	20	50
Demand	40	10	30	20

d How much petrol would you get from 100 tonnes of Arabian crude oil?

e What proportion of Arabian oil contains the less useful long-chain hydrocarbons?

f Your oil refining company has an order for 40 tonnes of petrol. How many tonnes of Arabian oil will you have to distil to get this?

g Your customer surveys suggest that for every 40 tonnes of petrol you sell, you will only be able to sell 20 tonnes of the long-chain fraction. How much long-chain fraction will remain unsold?

h Write a letter to your managing director, explaining why your company needs to build a cracking plant to overcome your supply and demand problems.

Questions

1 Copy and complete the following sentences. Choose from:

vaporised cracking catalyst long-chain smaller

Unwanted __________ hydrocarbons are chopped up into __________ and more useful molecules by __________. In this process, the oil is __________ and passed over a hot __________.

2 North Sea oil gives 25% petrol. Why do you think North Sea oil costs more than Arabian oil?

Summary

- Large hydrocarbon molecules may be broken down into smaller molecules by thermal decomposition.
- This process is called cracking.

6:6 Plastics

Plastics

Cracking makes more short-chain hydrocarbons for fuels. It also makes small molecules that can be used to make other useful materials.

These small molecules can be made to 'pop' together to make very long chains called **polymers**. Polymers made from hydrocarbons are also known as **plastics**.

Polymers are named by putting 'poly' in front of the name of the small molecule. 'Poly' means 'many'. The simplest molecule used when making polymers is called ethene, so the polymer made from this is called **poly(ethene)**. This is often shortened to polythene.

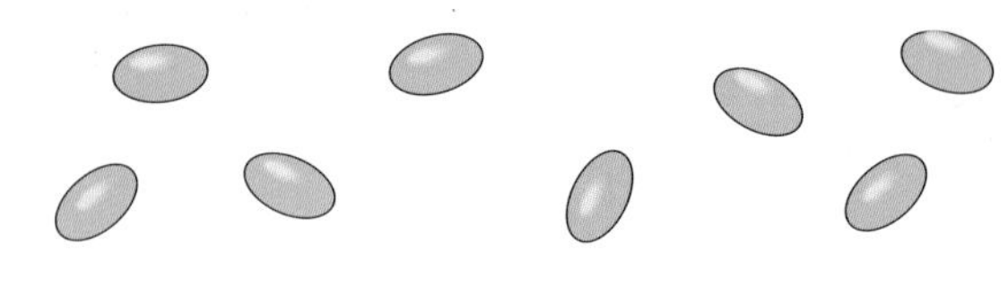

Cracking makes small ethene molecules

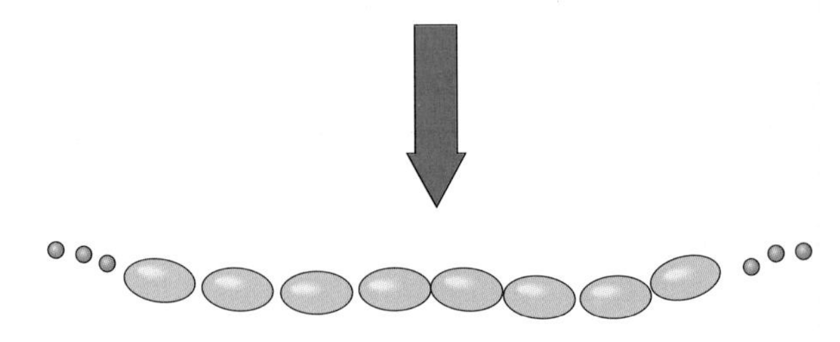

These can be made to pop together to form poly(ethene)

a Another small molecule is called propene. What would the plastic made from this be called?

Poly(ethene)

Poly(ethene) is a waxy solid that is very easy to shape. It can be moulded into bottles for drinks or even for dangerous chemicals such as bleach or acid. It can also be rolled into thin but tough, flexible and waterproof sheets. This is ideal as a food wrap or for plastic bags.

b What advantages do poly(ethene) bottles have over glass ones?

Poly(propene)

Poly(propene) is a tougher plastic than poly(ethene) and may be coloured easily too. It is used to make hardwearing things such as buckets, crates or even school chairs. It is also made into fibres for carpets and ropes.

c What polymer do you think washing up bowls are made from? Explain your answer.

The plastics for the job

Use this information table to answer the following questions.

Type of plastic	Properties	Uses
poly(ethene), PE	soft, flexible, looks waxy, lets some light through, melts at 80 °C, scratches easily	rubbish bags, squeezy bottles, dustbins
poly(propene), PP	hard but still flexible, looks waxy, melts at 145 °C, withstands solvents, very versatile material with many applications	drinking straws, microwave ware, plastic garden furniture, baby baths, plastic lunch boxes
poly(vinyl chloride), PVC	flexible, clear, elastic, hardwearing	garden hose, shoe soles, cable insulation, watch straps
poly(styrene), PS	rigid, brittle, opaque, semi-tough, melts at 95 °C, affected by fats and solvents	plastic cutlery, 'plastic' glasses, low-cost brittle toys, video cases, radiocassette player casings
expanded poly(styrene), EPS	foamed, lightweight, energy absorbing, heat insulating	foamed polystyrene hot drink cups, foamed meat trays, protective packaging for fragile items
PET	Clear, tough, solvent resistant, often used as a fibre	carbonated soft drink bottles, pillow and sleeping bag filling, textile fibres

d Which plastic is used for protective packaging?

e Which plastic is used for shoe soles? What properties make it suitable for this?

f Which plastic is used to make video cases and cheap toys? Why do these sometimes crack if they are dropped?

g Would you use poly(ethene) or poly(propene) to make a bowl for boiling water? Explain your answer.

h Why is expanded polystyrene used for hot drink cups?

i Which type of plastic bottle could be recycled to make sleeping bag filler?

Questions

1 Copy and complete these sentences. Choose from:

plastics **polymers** **molecules**

Some small __________ made by cracking can 'pop' together to make very long chains called __________. Polymers are also known as __________.

2 What is the name of the small molecule that is used to make poly(styrene)?

3 Rearrange these sentences to explain how poly(ethene) is made from crude oil.

- The ethene is made to polymerise.
- Crude oil is fractionally distilled.
- Ethene is separated out.
- Some of the long-chain fractions are cracked.

Summary

- Some small molecules made during cracking are used to make plastics (polymers).
- Poly(ethene) is used for plastic bags and bottles.
- Poly(propene) is used for crates and ropes.

6:7 Plastic problems

Isn't it great!

Plastics are wonderful, aren't they ...

- bottles for our soft drinks;
- cartons for our hamburgers;
- plastic trays and cling film for meat;
- plastic foam packaging;
- ... and don't forget plastic rubbish bags to throw it all away!

Use	Percentage
packaging	35
building	25
electrical	10
transport	7
home and leisure	13
other	10

How we use plastics.

a Draw a pie chart of plastic use, using the data in the table.

The downside

Disposing of plastics is, of course, the problem. Over a million tonnes of plastic packaging are produced every year in Britain. A lot of this ends up as litter, an ugly reminder of our wastefulness.

b Suggest two ways in which plastic rubbish gets onto a beach.

Nature's way

Paper is thrown away too, but this natural material rots down rapidly in the soil – it is **biodegradable**. Plastics, however, are not biodegradable, and may take tens or even hundreds of years to break down.

c People throw away banana skins and orange peel, too. Why is this less of a problem than crisp packets?

Most waste plastic does end up in the dustbin, where it forms about 20% of the rubbish. It then goes onto the local tip or landfill site. But landfill sites are filling up fast. How long will it be before we drown in the sea of our own rubbish?

d Suggest two other ways in which we could dispose of our rubbish.

Different people, different solutions

Earnest Green, environmental activist
"We are being buried alive by plastic that pollutes our environment. We must ban all plastic packaging now before it's too late. Some parts of New York State, USA, already have laws that have forced supermarkets to go back to using paper bags."

Jayne Middleton, local campaigner
"Plastics are produced from oil by processes that use large amounts of energy. What a waste to just throw it all away. The solution is to recycle wherever possible."

Susan Foresight, chemist
"Smarter technology can solve the problem. Chemical firms such as ICI are working on the development of new plastics which are biodegradable."

Jack Burnham, combustion engineer
"Plastic waste contains as much stored energy as fossil fuels. Why bother to recycle when you can just burn the plastic waste as fuel for a power station? Think how much coal and oil you could save."

What do you think?

Either:

Draw up a table with two columns and list the advantages and disadvantages of plastics rather than 'natural' materials.

Advantages	Disadvantages

Design a poster to make people think about this problem, using the points you have noted.

Or:

Working with a group, take on one of the characters quoted here. Try to develop their arguments by thinking the problems through and discussing them with your group.

You could try to find out more about the issues by visiting a library, searching the Internet or writing to environmental groups or industrial companies.

When you have gathered enough information, either write a report expanding on your ideas or hold a mock 'public inquiry' or debate on the way forward.

Summary

- Plastics can cause environmental problems because they are not biodegradeable.

6:8 The atmosphere

The **atmosphere** is a blanket of air wrapped around the planet Earth. Dry air is a mixture of two main gases: oxygen, the gas we need to breathe; and nitrogen, an unreactive gas. The proportions are approximately:

- one-fifth oxygen (20%);
- four-fifths nitrogen (80%);
- there are also small amounts of carbon dioxide and some unreactive gases.

The atmosphere has been like this for 200 million years.

a That's what's in *dry* air. But what other common chemical is in the air in this photograph?

The first atmosphere

When the Earth first formed, 4.5 billion years ago, things were very different. The Earth was very hot and was covered with volcanoes. These volcanoes blasted out gases to form the first atmosphere. This first atmosphere was mostly made of carbon dioxide and steam, with a little methane and ammonia.

b Near Naples in Italy there is a cave known as 'the mouth of hell' that fills up with volcanic gas. Any dogs that wander in die by suffocation. What gas causes this?

4.5 billion years ago.

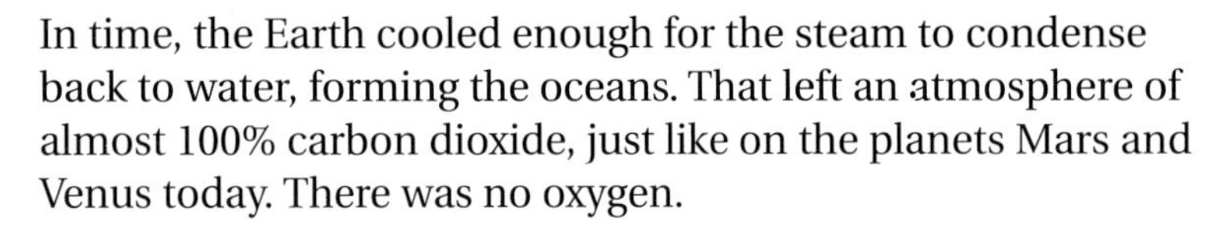

In time, the Earth cooled enough for the steam to condense back to water, forming the oceans. That left an atmosphere of almost 100% carbon dioxide, just like on the planets Mars and Venus today. There was no oxygen.

Simple forms of life evolved on Earth in this oxygen-free environment. Then about 3 billion years ago simple plants evolved and changed the world.

c Where can you find organisms that live without oxygen today?

3 billion years ago.

The first pollutant on Earth ...

Plants make oxygen as a waste gas of **photosynthesis**. To all the other life forms at the time, oxygen was a deadly poison! So the growth and spread of simple plants led to massive pollution of the world's oceans – with oxygen ...

By 2 billion years ago, other life forms were nearly wiped out. Just a few bacteria survived in oxygen-free deep ocean mud or stagnant pools.

d What happens to these primitive bacteria if you stir up the mud and expose them to oxygen?

2 billion years ago.

Our changing atmosphere

e How long ago did the carbon dioxide level drop to 50%?

f For roughly what proportion of the Earth's history has the oxygen level been as it is today? (5%, 20%, 50%)

g For approximately how many millions of years has there been more oxygen than carbon dioxide in the atmosphere?

h What major gas in the atmosphere is not shown on the graph?

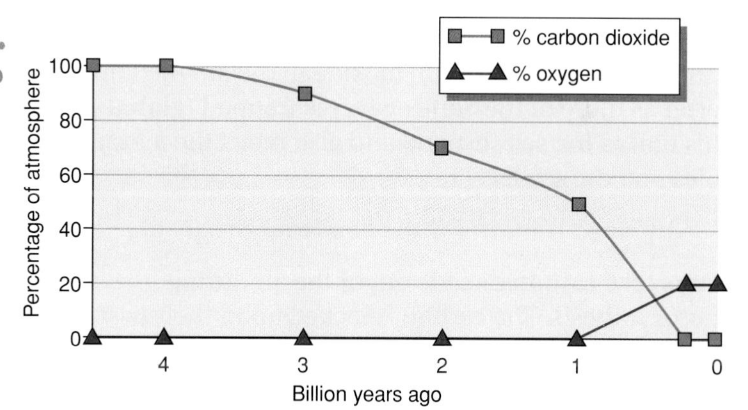

Questions

1 Copy and complete the following sentences. Choose from:

plants **polluted** **oxygen** **atmosphere**

The Earth's first __________ was mostly carbon dioxide, with no __________. Oxygen only appeared after __________ evolved. Oxygen __________ the oceans, killing off much of the early life.

2 Draw a pie chart of the amounts of oxygen and carbon dioxide in the air today.

3 Which planets still have atmospheres like the Earth's early atmosphere?

Summary

- The atmosphere is made of 20% oxygen and 80% nitrogen.
- The first atmosphere was mostly carbon dioxide.
- Oxygen is present in the atmosphere because of plants.
- The first oxygen killed off the more primitive organisms.

6:9 Keep it in balance

What do plants do for us?

Plants changed the Earth's atmosphere, killing off early life forms with toxic oxygen. But animal life has since evolved to rely on this same gas.

You couldn't live unless plants produced the oxygen you need to breathe. So plants are now the 'good guys' of the planet, keeping our atmosphere in good condition for us.

a What else do we need plants for, apart from oxygen?

Plants put oxygen into the air – but they also take out carbon dioxide. This is important, as the amount of carbon dioxide in the air helps to control the temperature of the Earth. The carbon dioxide is like a blanket, keeping the heat in. This is called the **greenhouse effect**.

If there is too much carbon dioxide in the air, the Earth gets hotter as more of the Sun's energy is trapped (**global warming**). This makes the seas expand and also melts the ice caps at the poles – so the sea level rises.

b Why does it matter if the sea level rises?

Plants take carbon dioxide out of the air during photosynthesis. The carbon is locked up in their tissues while they are alive. When they die, some is released but some stays trapped in the fossil fuels, coal, oil and gas.

$$\underset{\text{(free in the air)}}{\text{carbon dioxide}} + \text{water} \xrightarrow{\text{photosynthesis}} \underset{\text{(trapped carbon)}}{\text{glucose}} + \text{oxygen}$$

Plants use glucose to grow, so the carbon is trapped in their tissues.

c How do plants help to prevent global warming?

Shellfish, too

Plants have had help removing carbon dioxide from the atmosphere – from shellfish! Shellfish and corals build shells from calcium carbonate. They get this from calcium salts in sea water and carbon dioxide. Vast amounts of carbon dioxide are locked up in the limestone that has formed from the remains of shellfish and corals.

$$\underset{\text{(soluble)}}{\text{calcium salts}} + \text{carbon dioxide} \rightarrow \underset{\text{(insoluble)}}{\text{calcium carbonate}}$$

d Which type of sedimentary rock helps to prevent global warming?

What about the nitrogen?

As the amount of carbon dioxide in the air fell, it was steadily replaced by the nitrogen we have today. Some of this came from a chemical reaction between ammonia and the newly formed oxygen.

Ammonia contains nitrogen: NH_3.

ammonia + oxygen → nitrogen + water

e What atoms are there in a molecule of ammonia? How many of each type?

Looking after the atmosphere

The amount of carbon dioxide in the air is now very small – just 0.03% or so. Scientists can measure how this level has changed over the last 2000 years. Here is a graph of their results.

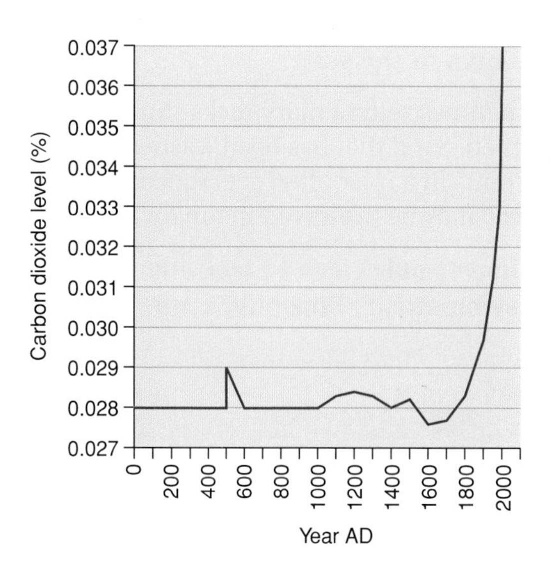

f Describe in words what the graph shows.

g There was a huge volcanic eruption in AD 500. What effect did this have on the carbon dioxide levels?

h How long did this effect last? (50, 200 or 500 years)

i How long has the recent rise in carbon dioxide levels been going on for?

j Do you think this could have been caused by volcanic activity? Explain your answer.

k What human activity over the last 200 years might have caused the recent rise in carbon dioxide levels?

l What do you think has happened to the average temperature of the Earth over the last 200 years?

Questions

1 Copy and complete the following sentences. Choose from:

Shellfish dioxide rocks limestone photosynthesis

Plants take carbon __________ out of the air during __________. __________ take carbon dioxide out of the air to make their shells. Some of this carbon remains trapped in __________ as fossil fuels and __________.

2 How did some of the nitrogen get into the air?

3 What effect might cement production have on the global temperature? (Hint: carbon dioxide is produced when cement is made from limestone.)

Summary

- Plants take carbon dioxide out of the air.
- Some is locked up for millions of years in fossil fuels.
- More carbon is locked up in limestone.
- Ammonia in the air reacted with oxygen to form nitrogen.

6:10 Evidence in the rocks

Sedimentary rocks

Rocks made from broken fragments are usually **sedimentary rocks**. They form when mountains are weathered and eroded away and the debris is washed into the sea. The fragments are then slowly squashed and cemented together to form new rock.

a What carries the sand and clay from eroded mountains down to the sea?

Sometimes sedimentary rocks show clues about how they were formed. Sand that has been washed by waves on a beach or currents in a river often gets rucked up into **ripple marks**. These may be preserved in the rock.

b River ripples lean to one side, beach ripples are symmetrical. Which type are the ripples shown below?

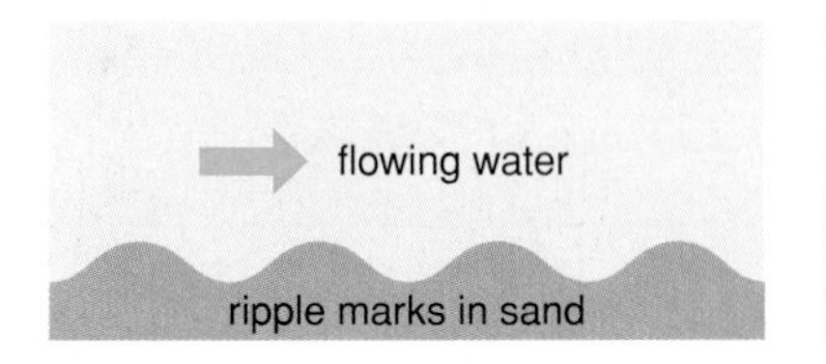

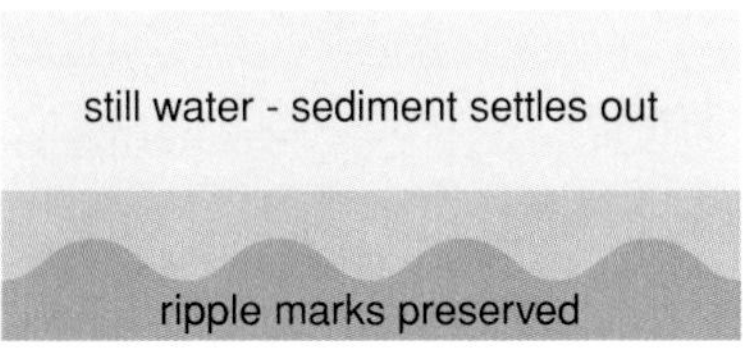

Sediment and the sea bed

Sometimes rivers flood and carry lots of sediment to the sea. At other times, the rivers flow more slowly and carry no sediment at all. In these clear periods, the sediment starts to harden, and sea creatures live on this new sea bed.

When the river floods again, the sea creatures may be buried and killed. Their shells are left as **fossils**.

In this way, sediments form in horizontal **beds**. If you split a sedimentary rock along the line of its bed, you might be lucky and find a fossil.

c Why do sediments form in horizontal beds? (What force is pulling them down flat?)

d Which are the youngest beds of sediment, those at the top or those at the bottom?

Stage 1

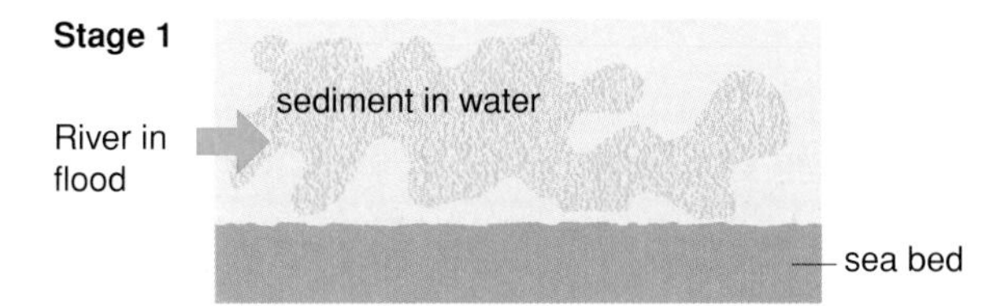

Stage 2

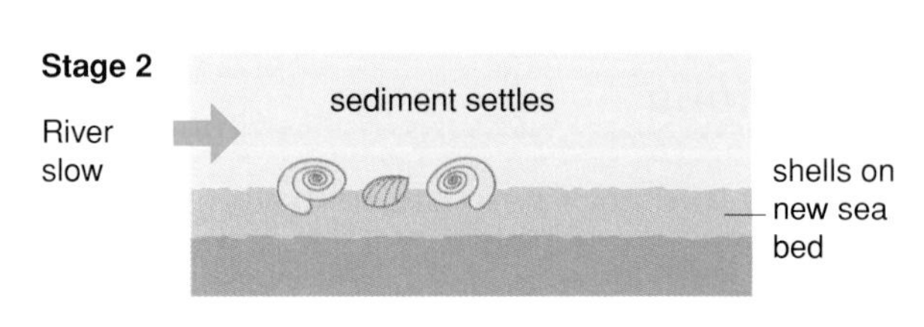

Stage 3

Igneous rocks

Igneous rocks form from molten rock (**magma**). They are made from crystals arranged randomly.

Granite is an igneous rock that has large crystals. This means it must have cooled slowly deep underground. When magma forces its way *into* the crust it forms **intrusive** igneous rock.

Lava from a volcano has tiny crystals because it cooled rapidly on the surface of the Earth. When magma pours *out* onto the surface it forms **extrusive** igneous rock.

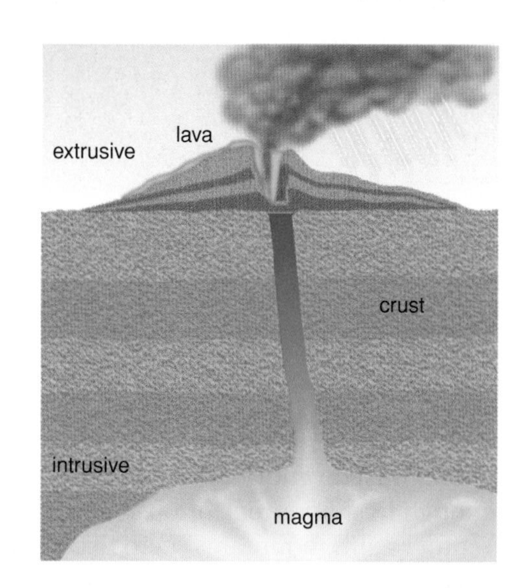

Metamorphic rock

If sediment is buried deep enough, the high temperature and pressure makes it change and recrystallise. Rocks that have changed like this are called **metamorphic rocks**. The crystals usually form in layers.

Limestone turns into marble when it is metamorphosed. Any fossils in the limestone are lost as the calcium carbonate recrystallises.

Limestone forms marble.

Full circle

Metamorphic rocks have been heated and changed while still solid. But if the temperature gets too high the rocks will melt and so become magma. This leads to new igneous rocks, linking the rock types together in the **rock cycle**.

e Describe how natural processes 'recycle' rocks.

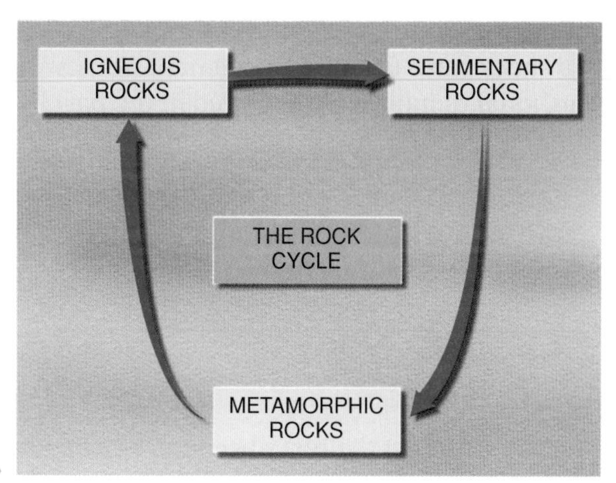

The rock cycle.

Questions

1 Copy and complete the following sentences. Choose from:

metamorphic **igneous** **Sedimentary** **magma**

__________ rocks form as old mountains wear away. If they are buried deep enough and heated they are changed to __________ rocks. If they are heated too much they melt and form __________. This can then form more __________ rock.

2 Patti found some igneous rock with large crystals clearly visible. She told her friend Jameisha that she had found some lava, but Jameisha disagreed, saying it must be from an igneous intrusion. Who was right? Explain your answer.

3 Some 'marble' tombstones have fossil shells visible in the rock. Could this really be marble? Explain your answer.

Summary

- Rocks contain evidence of how they were formed.
- Igneous, sedimentary and metamorphic rocks are linked in the rock cycle.

6:11 Active Earth

These rocks were laid down in horizontal beds at the bottom of the sea. Look at them now – uplifted, buckled and broken.

a Draw a sketch of this cliff face, picking out some of the buckled beds of sediment.

There must be enormous forces at work in the Earth. Big enough to shatter rock or push up towering mountains. Yet rocks and mountains seem fixed and everlasting.

Earth processes are usually very slow, taking millions of years to make big changes. Can we ever hope to catch a glimpse of these processes in action?

b What explosive evidence is there for Earth activity?

Stair Hole, Dorset

Earthquake!

Every few months there is news of a major **earthquake** somewhere in the world – Japan, Turkey or California, for example. The ground is shaken by an enormous shock wave that shatters buildings. Why do these occur?

When rocks near the surface of the Earth are squeezed by great forces, they may suddenly break and move. An earthquake is the shock wave from this movement. You can see where earthquakes must have occurred in the past by looking for large cracks in the rocks called **fault lines**.

c One earthquake may move the rocks just a few centimetres along a fault line. Some old fault lines show movement of tens or even hundreds of metres. How could this happen?

Deeper down

Deep in the crust the rocks become very hot – perhaps 500 °C or so. This makes the rock soft like Plasticine. If this soft rock is squeezed, the layers buckle and fold rather than break. The deeper and hotter the rocks are, the tighter the folding can get.

d There are places in Scotland where the rocks have been folded up and over like crumpled sheets. Did this happen near the surface or deep in the crust?

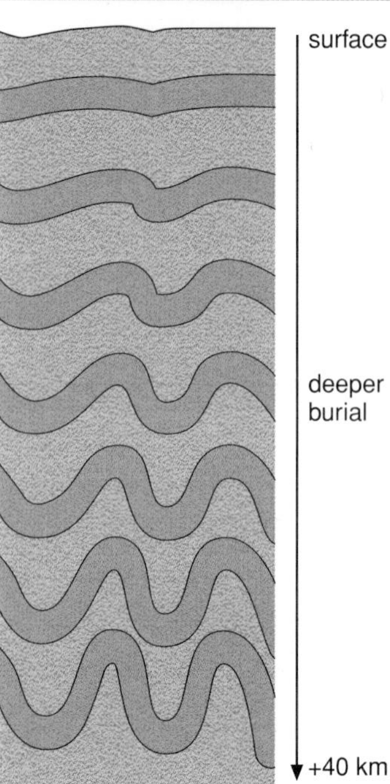

The deeper you go, the more intense the folding can be.

The mountain link

Sedimentary rocks form as old mountains wear away.

They are changed to metamorphic rock when they are deeply buried. They may even melt if they get hot enough.

This natural recycling of rocks is linked to the great movements of the Earth that cause faulting and folding.

Evidence of faulting, folding, metamorphism and igneous activity is found in all the mountain belts of the Earth. This shows that high temperatures and pressures are linked to the formation of mountains.

The rock cycle takes place over hundreds of millions of years as old mountains are worn away and new ones are forced up to take their place.

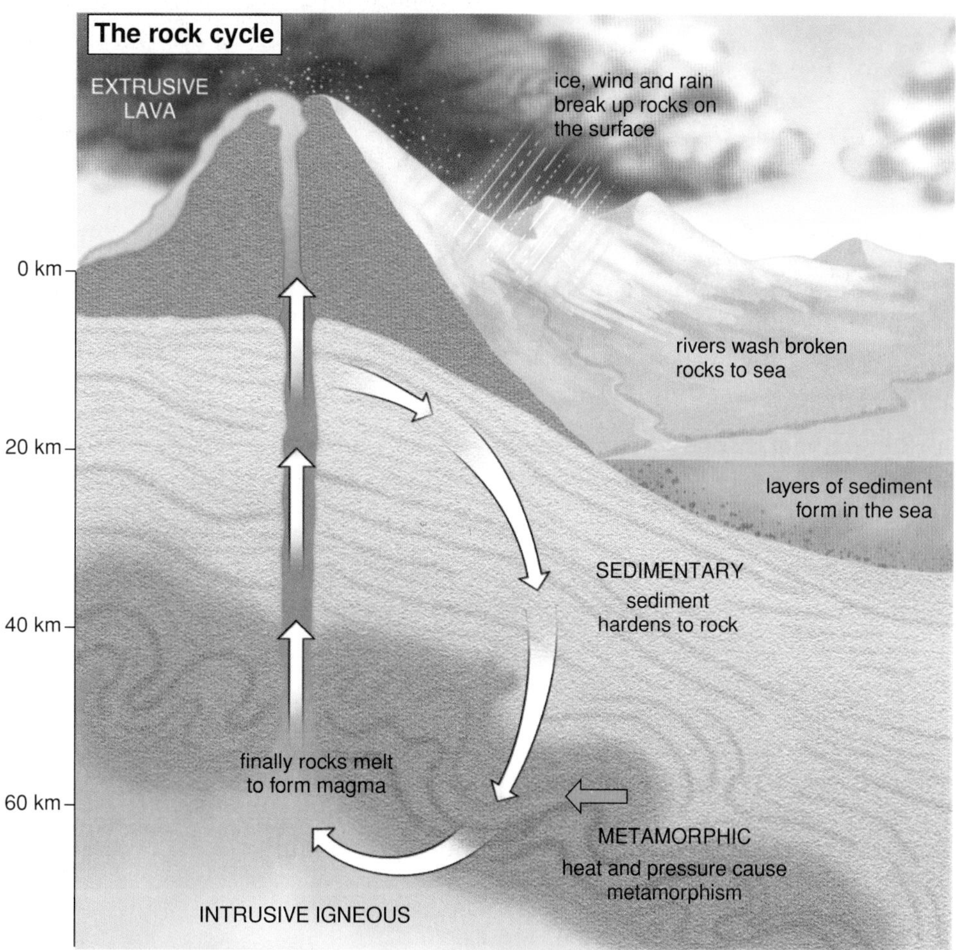

e Limestone is often faulted, marble is often tightly folded. Explain why this might be.

f Sedimentary rocks form as a result of the destruction of old mountains. Use the diagram to help you describe how they also provide material to build new ones!

Questions

1 Copy and complete the following sentences. Choose from:

earthquake fault folded squeezed

Cold rocks that are __________ by great forces may break along a __________ line. This sets off an __________. Deeper, hotter rocks may be __________ instead.

2 If an earthquake occurred every 10 years on average, and the rock moved just 10 cm each time, how far would the rock have moved in a million years? Take it one step at a time:

- In 100 years it would move __________ m.
- In 1 000 years it would move __________ m.
- In 1 000 000 years it would move __________ m.

Summary

- The Earth is very active. Rocks may be uplifted, faulted, folded or even overturned.
- Rocks buried deep in the crust get very hot.
- Metamorphic rocks are linked to the formation of new mountains.

6:12 Earth structure

Moving continents

New mountains are pushed up and old ones are worn away. If that's not enough, the continents are moving around too! Europe is moving away from America at a speed of 2 cm every year ...

a **2 cm a year doesn't sound much, but how far would Europe move in 50 years, 50 thousand years and 50 million years?**

b **Look at the shape of the continents on both sides of the Atlantic Ocean. What does that suggest?**

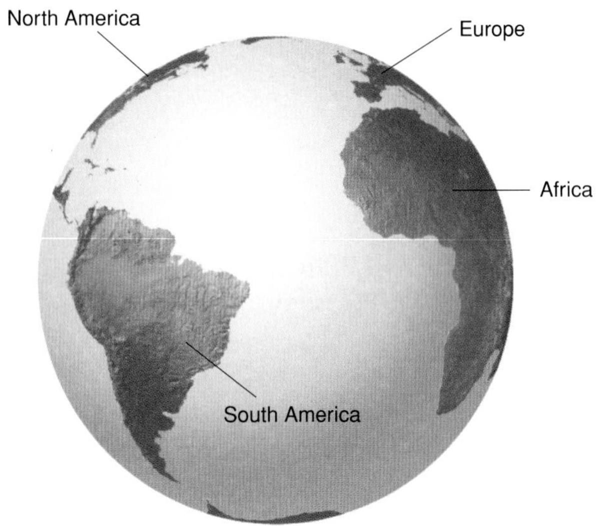

How can it happen?

We live on the surface of the Earth and our deepest mines barely scratch the surface. We call the rocky, surface part the **crust**, and that's just what it is – a thin hard layer above something that's very much hotter and softer.

c **It is 6500 km to the centre of the Earth. Our deepest mines go down 13 km. What fraction of the way to the centre is that? 1/500, 1/100 or 1/50?**

Inside, the Earth looks like a Scotch egg, with layers of different material one inside the other. Movement in these layers makes the crust move about.

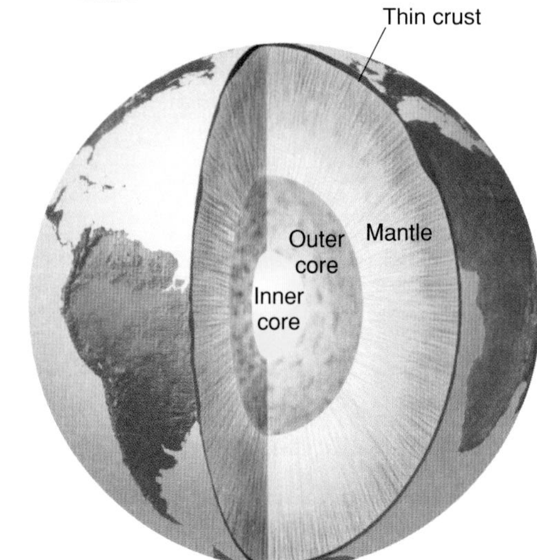

The mantle

The **mantle**, a hot rocky layer, extends almost halfway to the centre of the Earth. It acts like a solid in many ways but, over millions of years, it can move and flow.

d **In the 'Scotch egg' model, which would be the sausagemeat and which would be the egg?**

The core

- The **core** of the Earth is not made of rock but iron and nickel.
- The **outer core** is so hot that the metal is molten.
- The **inner core** is hot too, but the pressure is so great that it is solid metal.

How do we know?

We can only see the top of the crust, so how do we know what the Earth is like inside? There are many clues. Consider, for example, the density of the Earth.

The rocks of the crust are generally less than three times as dense as water, but the Earth as a whole is almost twice as dense as this.

Layers of the Earth	Density ($g\ cm^{-3}$)	Thickness of layer (km) (approximate)
crust	2.75	40
mantle	4.5	3000
outer core	8	2000
inner core	10	1500 (to centre)
Earth as a whole	5.5	13 000 (diameter)

e Draw a bar chart showing the densities of Earth's layers.

Space invaders

Some meteorites are fragments of an old planet between Mars and Jupiter that broke up. There are two types of meteorites, rocky ones and nickel/iron ones.

f How do meteorites support our understanding of Earth's mantle and core?

g Doctors use sound waves to take 'ultrasound' images of babies. How is this like the earthquake shock waves 'X-ray' of the Earth?

Did you know?

Scientists use the shock waves from earthquakes to get an 'X-ray' picture of the Earth. They measure these on machines called seismometers that are placed all over the globe. The results are analysed by computer. You will find out more about this in the 'Waves' module.

Questions

1 Copy and complete the following sentences. Choose from:

iron mantle core layered crust

The Earth has a __________ structure. We live on a very thin __________. Beneath this is the rocky __________. In the middle is the molten __________ /nickel __________, with a solid centre.

2 a Draw a scale model of the Earth using the layer thicknesses given in the table above. Use a scale of 1 cm for every 1000 km. The crust will be a very thin pencil line!

b Label your drawing and add a phrase to describe each layer.

3 The crust is solid, the outer core liquid. What is odd about the mantle?

Summary

The Earth has a layered structure with:

- a thin rocky crust;
- a soft rocky mantle;
- a liquid nickel/iron core with a solid centre.

Overall, the Earth is twice as dense as the rocks of the crust.

6:13 Tectonic plates

Why are there no volcanoes in Britain?

Earthquakes and volcanoes are common enough – but not in Britain. If you plot them on a map they occur in clear belts across the globe:

- the 'ring of fire' around the Pacific;
- the young mountains from the Alps to the Himalayas;
- the middle of the Atlantic and Pacific Oceans.

The reason for this pattern is simple enough. The crust of the Earth is broken into jigsaw pieces called **tectonic plates** that are moving about. The plates themselves are quite rigid as the crust is attached to a slab of upper mantle. This thicker, double layer of crust and upper mantle is called the **lithosphere**.

The earthquakes and volcanoes occur at the *edges* of the plates, where the different plates meet.

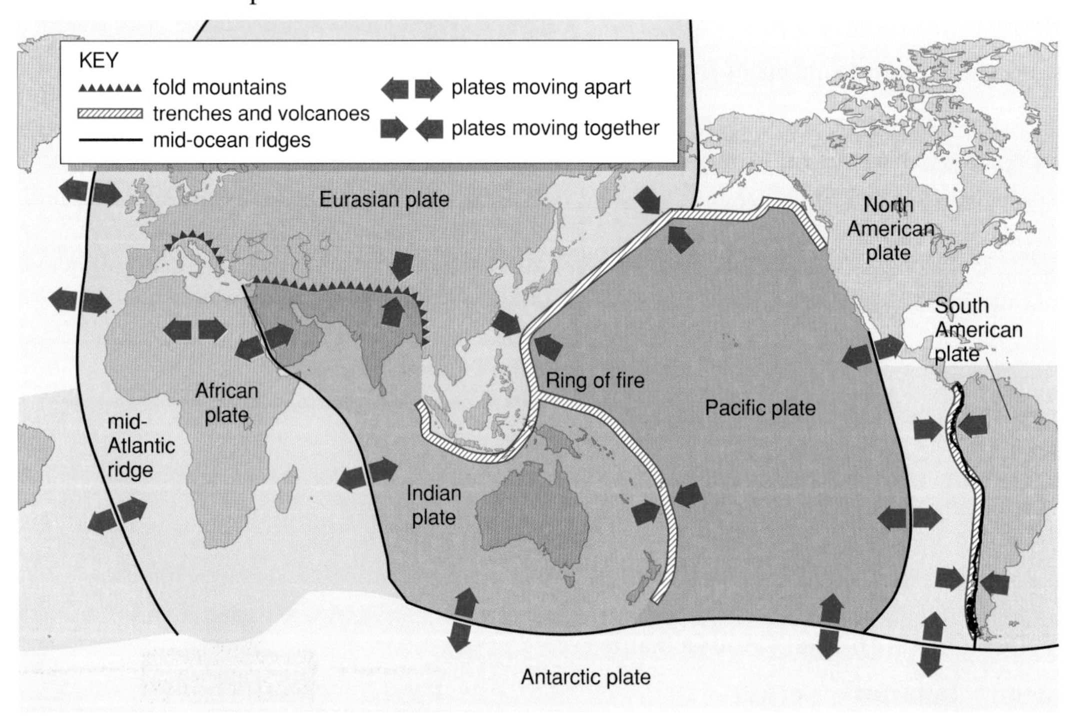

This map shows the main plates and how they are moving.

a From the map, what are the closest places to Britain where powerful earthquakes are common?

b Iceland is a volcanic island in the middle of the Atlantic Ocean. Why is Iceland getting wider?

c Which plate is Australia on?

d The Indian and Eurasian plates are moving together. Which famous mountains have been squeezed up between them?

Why do plates move?

If you heat a liquid from below, warm currents rise up and make the liquid swirl and mix. These are called **convection currents**.

The rocks in the mantle and core are being heated by natural radioactivity. Convection currents are 'swirling' through the mantle, too. It is just that they are moving very slowly – at just a few centimetres a year.

Given the very many millions of years of Earth history, however, slow convection currents can make continents move across the Earth and force new mountains high into the air.

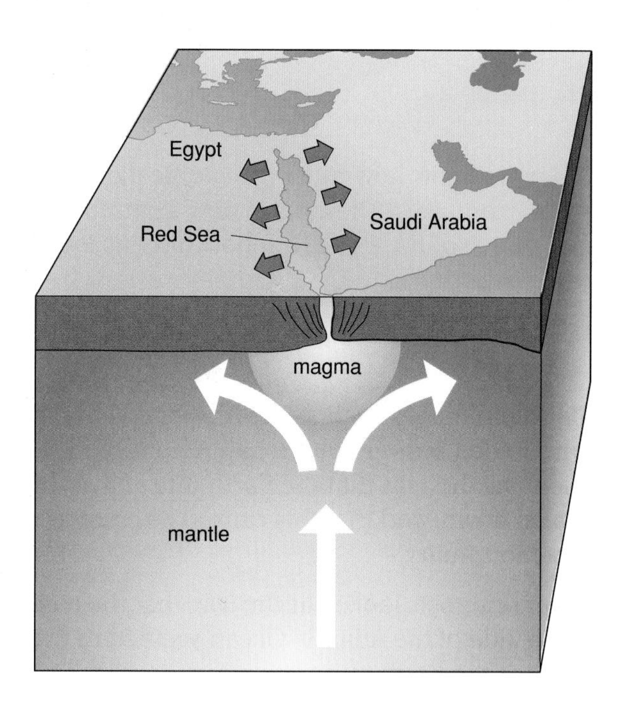

e **A convection current in the mantle is rising beneath the Red Sea and moving the African and Indian plates apart. What will happen to the Red Sea over the next few million years?**

f **The continents seem to be moving apart now, but the Earth is round! What might happen in another 250 million years?**

Did you know?

Most of the continents now seem to be moving apart, riding on their plates. This means that if you go back in time, they must have been closer together.

Go back 250 million years and the world looked very different. All the continents were stuck together in a supercontinent called Pangaea.

Questions

1 **Copy and complete the following sentences: Choose from:**

convection plates mantle edges volcanoes

The crust is broken up into tectonic __________. Earthquakes and __________ only occur along the __________ of these plates. The plates move slowly due to __________ currents in the __________.

2 **What is providing the heat energy to drive the convection currents in the mantle?**

Summary

- The crust is broken up into slabs called tectonic plates.
- Volcanoes and earthquakes occur at the plate edges.
- The plates move around because of convection currents in the mantle, caused by natural radioactivity.

6:14 Understanding the Earth

Today we know how moving tectonic plates cause volcanoes and earthquakes. These form new mountains and shape the surface of the Earth. But how did this understanding come about?

Early ideas

Mountain ranges such as the Alps and Himalayas are built from folded sedimentary rocks. A hundred years ago, some scientists thought that the Earth must have shrunk as it cooled down, and that this caused its crust to wrinkle up into mountains.

Other scientists looked at the way that the continents on either side of the Atlantic Ocean seemed to fit together. Perhaps the Earth was getting bigger, and this had torn the continents apart!

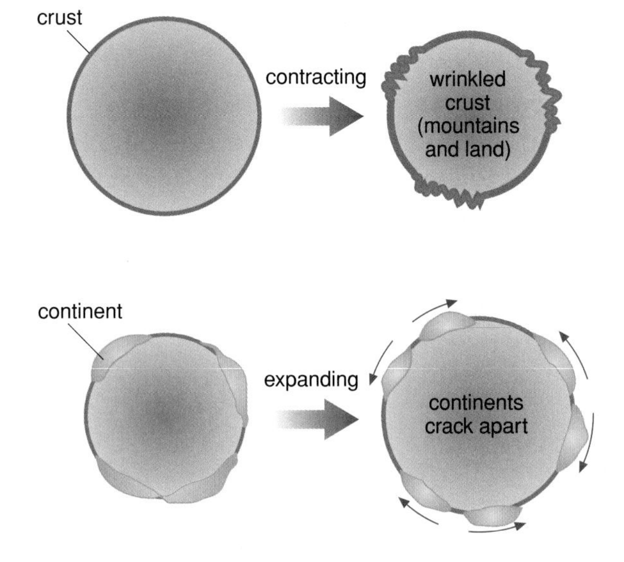

Some scientists believed that the Earth was contracting whilst others believed that it was expanding.

a Why couldn't both these ideas be correct?

In 1915, a scientist called Alfred Wegener looked in detail at the fit between South America and Africa and was amazed at how good it was:

- the different rock types matched up;
- the same fossils were found on both sides;
- there was evidence of an ancient shoreline that matched.

He concluded that they must have split apart, like matching halves of a torn picture. He called this **continental drift**, but he couldn't explain how it could work, so nobody believed him.

b Why didn't scientists believe Wegener?

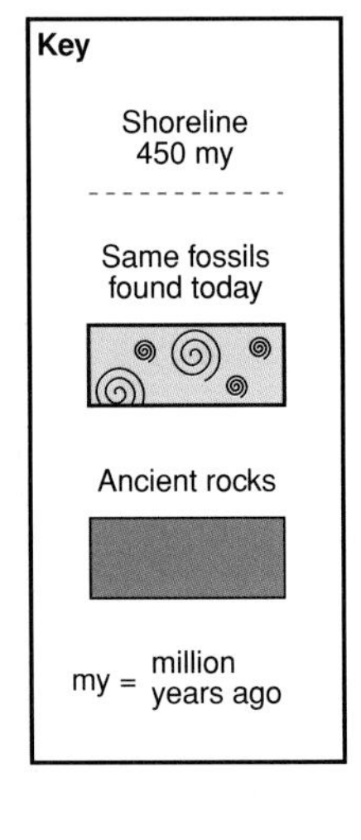

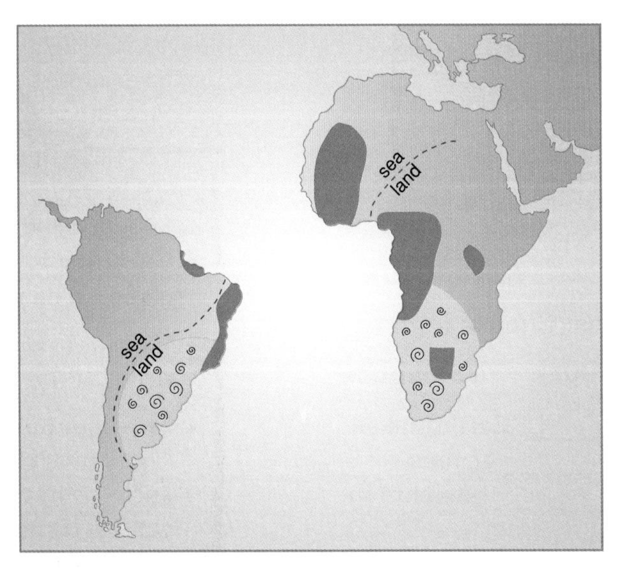

How the continents are today.

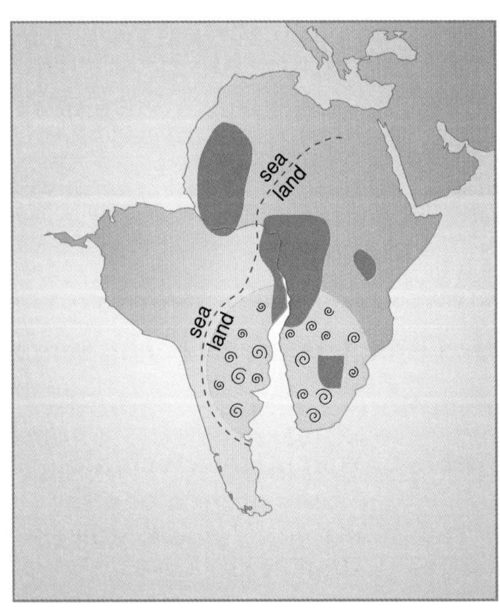

How Wegener thought the land must have been millions of years ago.

Unexpected mountains

In the 1950s, scientists started to survey the floor of the Atlantic Ocean.

The first, startling, discovery was a chain of undersea volcanic mountains running along its centre! No one expected there to be mountains at the bottom of the sea.

The scientists also found that all the rocks of the ocean floor had come from volcanoes. The surprising discovery was that the rocks got older and older as you moved away from the central ridge.

The ridge of volcanoes was running along a great crack in the crust! Magma from the mantle pushed into the crack, cooled and set. This new rock then cracked, and so on …

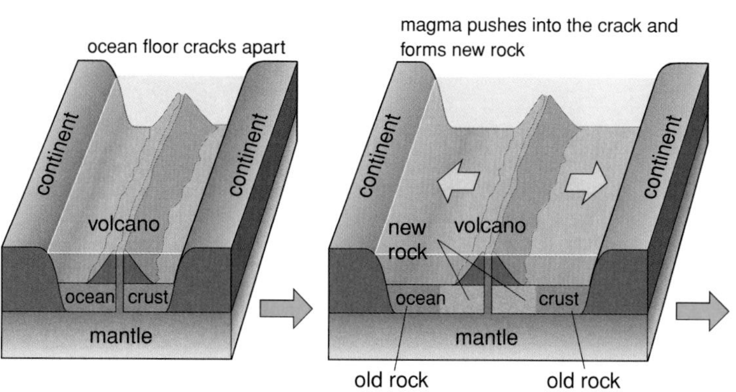

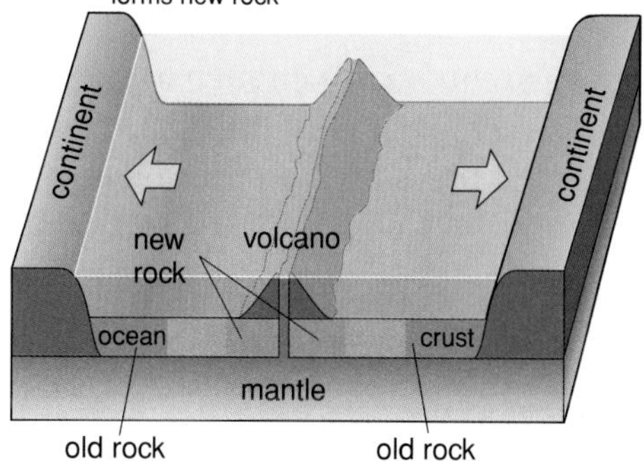

Wegener was right – the continents are drifting apart.

c How is new rock made at the bottom of the ocean?

d Why does the centre of the sea floor keep cracking open?

e How does this support continental drift?

Summary

- A hundred years ago scientists could not explain why mountains, continents and oceans formed.
- Wegener looked at the jigsaw fit of the continents and suggested they had drifted apart.
- He was not believed because he could not explain how this could work.
- 50 years later, new evidence was discovered that supported Wegener and led to the theory of plate tectonics.

Questions

1 Copy and complete the following sentences. Choose from:

explain plate drifted continents

Wegener explained the jigsaw fit of the __________ by suggesting that they had __________ apart. Nobody believed him because he could not __________ how it worked. Later discoveries supported him and led to the __________ tectonic theory.

2 Why would you expect to find the same fossils in land that used to be joined millions of years ago?

3 What two things match perfectly when Africa and South America are pushed back together?

4 Draw up a timeline for the development of the continental drift/plate tectonics theories, marking in the key discoveries.

6:15 Predict and survive?

The great volcanic cone of Vesuvius looms over the City of Naples in Italy. Nearly 2000 years ago, a gigantic eruption sent a cloud of red-hot gas down on the thriving city of Pompeii, killing all its 5000 inhabitants. Should the million people who live in Naples today be worried? You'd better believe it!

A killer on the doorstep.

Vesuvius last erupted in 1944, when lava stopped just on the edge of the city. It has been sleeping since then, but:

- Vesuvius has a long history of eruptions;
- it will erupt again;
- the only question is 'when'.

If the people of Naples are to have a chance of survival, they need to have advance warning so that they can evacuate the city. Scientists must try to predict when the volcano will erupt, and that is far from easy. But there are some warning signs.

a What was different about the 1944 eruption, compared to the one that destroyed Pompeii?

Warning signs

Increased temperature
There is a great chamber full of molten magma beneath Vesuvius, and hot gases from this escape into the crater. The temperature of this gas is about 90 °C – hot enough to cook eggs. If that temperature rises, it might mean an eruption is coming soon.

More earthquakes
You get a lot of small earthquakes around Vesuvius. These earthquakes get stronger and more common before an eruption. Hundreds of small earthquakes have occurred in the last few years – the most active period since the last eruption.

More gas
The gas is mostly carbon dioxide and toxic, foul-smelling hydrogen sulphide. You get more gas before an eruption.

Rising land
As magma pushes in below the volcano, the ground rises. A harbour near Vesuvius has risen nearly 4 m out of the sea in the last 30 years.

b List the four things to look out for before an eruption.

c **La Sulfatara is a small crater to the west of Naples. Sulphurous gas is escaping at 165 °C. The region has had lots of small earthquakes recently and the crater floor has risen by a few metres. There is a large housing estate built on the rim of the crater. Write a letter to the local residents' association, warning them of the dangers.**

When to evacuate?

The problem is that all of these features show that an eruption is on its way, but they might start to occur months or even years before the event. There is no way that scientists can say *exactly* when an eruption will occur.

In 1985, an Italian scientist spotted these early warning signs around a volcano in Colombia and proposed that thousands of people should be evacuated from the nearby valleys. The advice was still being studied when the volcano erupted, killing thousands of people.

d **How quick or easy do you think it would be to organise the evacuation of a million people from Naples? Where would they go?**

How big is the problem?

Some volcanoes have bubbling lava in their craters, which simply spills over the edge during eruptions. Vesuvius is far more dangerous – the lava has set to form a plug in the crater neck. The pressure builds up and up inside the volcano, until the plug 'blows'.

The longer it is between eruptions, the bigger the bang when the volcano blows. When Vesuvius last erupted, in 1944, the pressure had being building for just 15 years ...

Year of eruption	Years since previous eruption	Millions of cubic metres of lava erupted
1794	34	27
1858	64	120
1872	14	20
1906	34	80
1929	23	12
1944	15	25
20__?		

Question

1 **Look at the data table above for Mount Vesuvius.**

a **Plot a graph of the amount of lava against the time between eruptions.**

b **What pattern does this graph show?**

c **Draw a line of best fit.**

d **Use this line to predict the volume of lava that would be produced if Vesuvius erupted today.**

e **Given where the 1944 lava stopped, should the people of Naples be worried? Explain your answer.**

Summary

- Scientists try to predict when volcanic eruptions will occur in order to save lives.
- Although there are many warning signs, it is still impossible to say exactly when an eruption will occur.

6:16 Predicting earthquakes

The effect of a magnitude 7 earthquake.

In August 1999, a terrible earthquake struck Izmit in Turkey, killing more than 40 000 people. The earthquake, which measured 7 on the Richter scale, struck without warning. The people died as their homes were shaken to rubble in seconds. What could have been done to prevent this terrible disaster?

The table shows the Richter scale for earthquakes.

Number	Effects
5	chimney pots fall – the worst we get in Britain
6	houses badly damaged
7	most buildings collapse – bridges and roads destroyed
8	total destruction – ground rises and falls

An earthquake of magnitude 6 is *ten times stronger* than an earthquake of magnitude 5, and so on.

a What happens when a magnitude 7 earthquake strikes?

How often do powerful earthquakes occur?

Many powerful earthquakes happen every year. Most of these strike uninhabited regions, and so cause few problems. Even so, earthquake disasters do happen with sad regularity.

b Why are some of the most powerful earthquakes not reported on the news?

Izmit was the worst earthquake in Europe since 1939, when 100 000 people died at Messina in Sicily. But when a colossal magnitude 8 earthquake hit Tangshan in China in 1976, the death toll reached 600 000.

In the twentieth century, nearly 1 million people died as a result of earthquakes. Earthquakes are by far the most deadly type of natural disaster.

Where do powerful earthquakes occur?

This is the easy part. Powerful earthquakes only occur along plate boundaries – that's why Britain never gets them. Destructive earthquakes occur where the plates are pushing together or sliding sideways, for example:

- along the 'ring of fire' around the Pacific;
- along the Alps/Himalayas belt of new fold mountains;
- zones such as the San Andreas fault zone in California.

Turkey is in the mountain belt. It is being squeezed between the African and Eurasian plates. It is not surprising that Turkey suffers from terrible earthquakes.

c What other great city is at risk in this 'collision zone'?

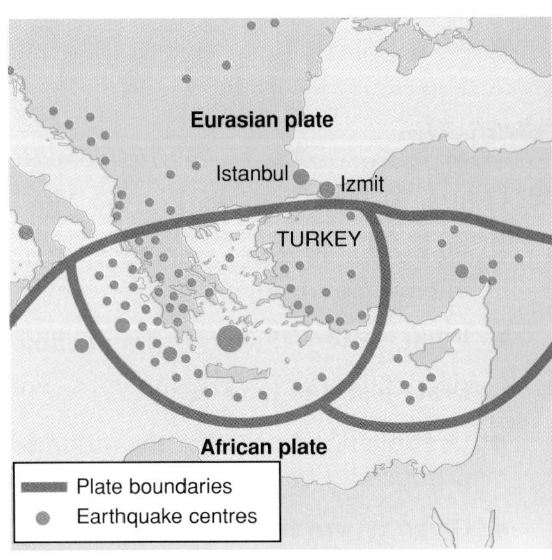

Why Izmit was at risk.

When will powerful earthquakes occur?

This is the difficult part! Local folk-tales often talk of animals behaving strangely just before an earthquake: dogs start to bark or all the birds go silent. Surely science can do better than that. Here are some of the ideas.

- Pre-shocks. Minor shocks often occur before a big shock. Perhaps these are what make the animals behave strangely.
- Changes in the water level in wells. Water levels often seem to drop in the period before an earthquake.
- Electrical changes. These often occur before an earthquake, but are very variable.

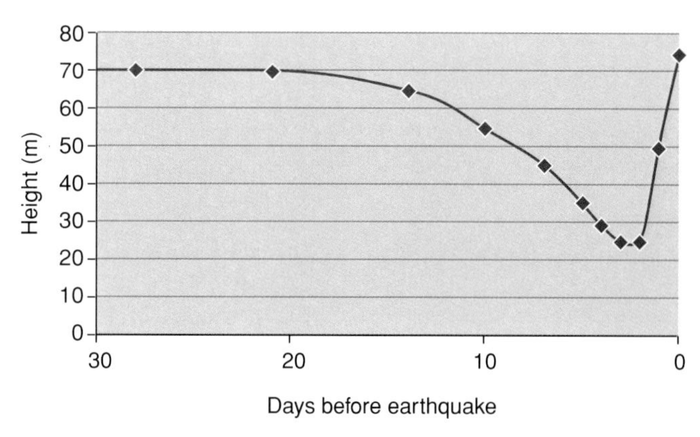

Water levels in a well before an earthquake.

The problem is that these effects may occur only hours or possibly days, weeks or months before the earthquake. So you may know that an earthquake is coming, but you still do not know *exactly* when.

The best way to save lives is to evacuate buildings and get people to camp out in the open. But do this too soon and people will get bored and drift back to their homes ... Perhaps people are still just as well off watching the animals!

Did you know?
The electrical effects you get with earthquakes might cause the strange lights that people mistake for UFOs!

d What do some animals appear to do before an earthquake? What might make them behave like this?

e What is the best thing to do if you think there is about to be an earthquake?

What can we do?

Some scientists now think that we will never be able to say precisely when an earthquake will occur. They say we would be far better off spending the money on ways to reduce the effects of earthquakes.

For example, building regulations could be strictly enforced in earthquake zones. Brick buildings that shake apart in an earthquake could be replaced by girder-framed buildings that sway but stay intact!

This tower in San Francisco, USA, was built to be earthquake-proof.

Question

1 The graph shows water levels in a well in the days before an earthquake.

a What pattern does this graph show?

b What happened *just* before the earthquake?

c Why are water levels studied in earthquake zones?

Summary

- Earthquakes are the deadliest kind of natural disaster.
- We know where earthquakes will occur but we may never be able to tell exactly when they will strike.

End of module questions

1 These questions are about crude oil. Use the words below to complete the sentences.

hydrocarbons

hydrogen

chains

mixture

1.1 Crude oil is a __________ of different compounds.

1.2 The compounds in crude oil contain the elements carbon and __________.

1.3 The compounds in crude oil are called __________.

1.4 The carbon atoms form __________ in these compounds.

2 Plastics (polymers) such as poly(ethene) are made from crude oil. Four stages in the process are the following:

i very long-chain molecules (polymers) are formed

ii small molecules are formed

iii large molecules are cracked

iv small molecules are made to 'pop' together

Which of these orders is correct?

A iv → ii → i → iii
B iii → ii → iv → i
C ii → i → iv → iii
D i → ii → iii → iv

3 These questions are about sediments and sedimentary rocks. Use the words below to complete the sentences.

horizontal

fossils

sediment

weather

3.1 Rocks __________ and erode and the debris is washed into rivers.

3.2 Fast-moving rivers bring __________ to the sea.

3.3 This settles out to form __________ beds.

3.4 Shells that are buried may be preserved as __________.

4 These questions are about how sedimentary rocks are changed when they are deeply buried and squeezed. Use the words below to complete the sentences.

minerals

temperature

metamorphic

folds

4.1 When sedimentary rocks are deeply buried they undergo very high __________ and pressure.

4.2 The beds of rock can be bent up into __________.

4.3 The rocks are changed as new __________ grow.

4.4 The rocks change from sedimentary to __________.

5 Crude oil is split into fractions by distillation. Choose from the phrases below to label the diagram.

short-chain hydrocarbons out

crude oil vaporised

viscous oil

the vapour rises and cools

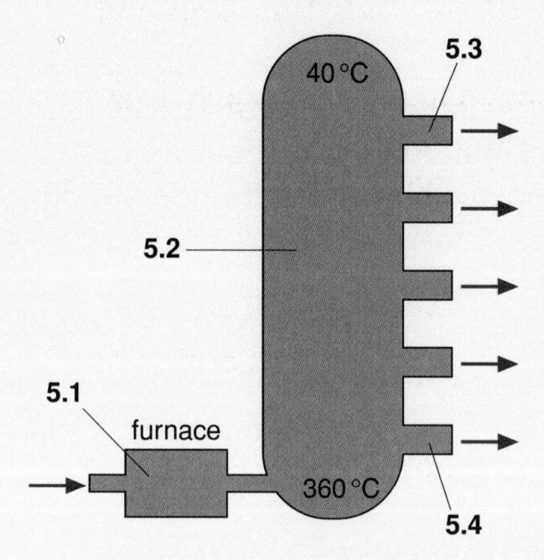

6 Choose the **two** best answers for questions 6.1 and 6.2.

6.1 Petrol and lubricating oil contain hydrocarbons. Petrol is runny and catches fire (ignites) easily. Lubricating oil is viscous and does not ignite easily.

A Petrol is less volatile than lubricating oil.
B Petrol hydrocarbons have shorter chains than lubricating oil hydrocarbons.
C Petrol has bigger molecules than lubricating oil.
D Lubricating oil is a better fuel than petrol.
E Lubricating oil has a lower boiling point than petrol.

6.2 The diagram shows a section through the Earth.

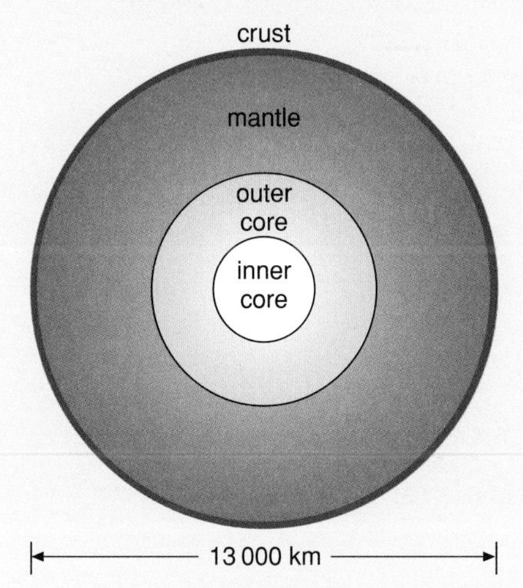

A The mantle is 10 000 km thick.
B The mantle is made from iron and nickel.
C The core as a whole has just over half the diameter of the Earth.
D The core is made from iron and nickel.
E The inner core is made of liquid rock.

7 Limestone is a natural form of calcium carbonate ($CaCO_3$). It is used to make cement and slaked lime.

7.1 A calcium carbonate particle is made up of ...

A one calcium atom, one carbon atom and one oxygen atom.
B three calcium atoms, three carbon atoms and three oxygen atoms.
C one calcium atom, one carbon atom and three oxygen atoms.
D one calcium atom, three carbon atoms and one oxygen atom.

7.2 When calcium carbonate is heated a gas is driven off. This gas is ...

A carbon monoxide.
B carbon dioxide.
C oxygen.
D ozone.

7.3 Breaking up a compound by heating is called ...

A double decomposition.
B thermal decomposition.
C neutralisation.
D polymerisation.

7.4 Cement is made by ...

A heating limestone and then adding water.
B heating limestone with clay.
C dissolving limestone in acid.
D mixing limestone with slaked lime.

8 Granite and basalt are both igneous rocks. Granite has large crystals, but in basalt the crystals are tiny and you need to use a microscope to see them.

8.1 How are the crystals arranged?

A Granite has random crystals but in basalt they are in layers.
B Basalt has random crystals but in granite they are in layers.
C In both granite and basalt the crystals are arranged in layers.
D In both granite and basalt the crystals are arranged randomly.

8.2 Where are the rocks formed?

A Basalt poured out onto the surface of the Earth but granite formed deep underground.
B Granite poured out onto the surface of the Earth but basalt formed deep underground.
C Granite and basalt both poured out of volcanoes as lava.
D Granite and basalt both solidified deep underground.

8.3 Both granite and basalt can be used for kerbstones. This is mainly because ...

A they are soft and easy to cut out into blocks.
B they are light and easy to transport.
C their tightly interlocking crystals mean that they are hard rocks.
D they look very attractive when cut.

8.4 You can find a lot of granite in Cornwall because ...

A there is a lot of volcanic activity in Cornwall today.
B it is a well-known tourist area.
C a lot of magma set in the crust there a long time ago.
D there is not much weathering and erosion in Cornwall today.

9 These questions are about the atmosphere and how it has changed over the history of the Earth.

9.1 The composition of the air today is approximately ...

A 20% nitrogen and 80% oxygen.
B 20% carbon dioxide and 80% oxygen.
C 20% oxygen and 80% carbon dioxide.
D 20% oxygen and 80% nitrogen.

9.2 4 billion years ago the Earth's atmosphere was mostly made from ...

A nitrogen.
B oxygen.
C methane.
D carbon dioxide.

9.3 There is oxygen in the atmosphere because ...

A animals need oxygen to breathe.
B animals make oxygen when they respire.
C methane reacts with carbon dioxide.
D plants make oxygen during photosynthesis.

9.4 There is less carbon dioxide in the atmosphere on Earth than there is on Venus because ...

A we burn fossil fuels on Earth.
B lots of carbon is locked up in limestone and fossil fuels.
C animals breathe out carbon dioxide.
D carbon dioxide has turned into nitrogen.

Module 9 – Energy

The idea of energy is a very important one in science. It can help us to understand what is going on in many different situations, for example:

- when a carload of people travels round a rollercoaster ride;
- when a creature eats food and runs around;
- when a chemical reaction happens;
- when a television uses electricity to show the latest music videos.

In this module, we will see how understanding the ways in which energy is transferred from a house can help you to save money. We will also see how understanding the energy transfers involved in generating and using electricity can help us to make better use of the Earth's natural resources.

Before you start this module, check that you can recall the answers to the following questions about energy:

1. What name is given to the energy of a moving object?
2. What materials are good conductors of heat energy?
3. Name three fossil fuels.
4. Is wood a renewable or non-renewable energy resource?
5. Where is most of our electrical energy generated?

9:1 Keeping heat energy in

In the winter, it is cold outside. If you touch a window pane, it feels cold. Heat energy from inside your warm house is being **conducted** away through the glass of the window.

In the UK, we have to keep our homes heated for much of the year. It is cold outside, and we want to stay warm. Schools, shops, offices and other buildings must also be heated.

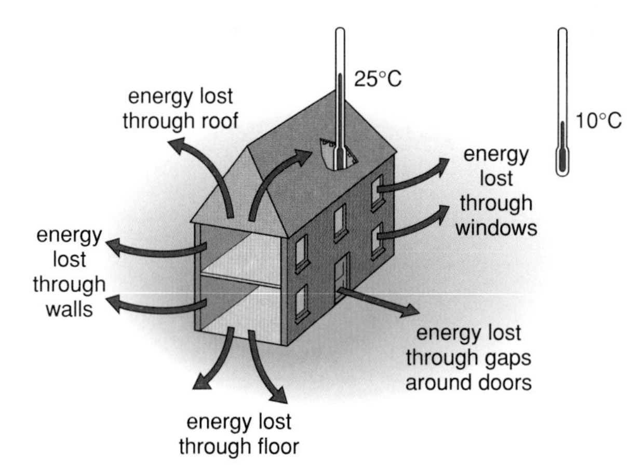

Energy is escaping from this house, because it is warmer than its surroundings. Energy is transferred from hotter places to colder places.

Conducting away

Heat energy from the warm house is conducted through the walls, roof and windows into the cold air outside. Heat is also conducted down through the floor into the cold ground.

The walls of some houses are made of a single layer of bricks. Heat energy conducts easily through the bricks.

Many houses have walls made of two layers of bricks, with a gap in between. The gap or cavity contains air. Air is very bad at conducting, so heat is transferred more slowly through the cavity wall. We call a material that does not conduct well an **insulator**.

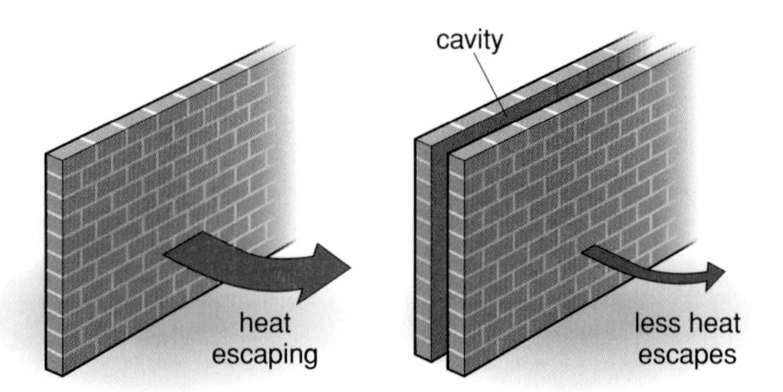

a Why does fitting thick carpets in the downstairs rooms of a house help to keep it warm?

Materials that insulate

An Inuit family may live in a house made of ice. The ice is cold, but the air inside is warm. The people stay warm because heat energy cannot escape easily through the ice. Their clothes, made from animal skins and furs, also help to keep them warm.

b How can you tell from the photo that ice is a good insulator?

Double glazing

A lot of heat energy is conducted out through windows. The panes of glass are much thinner than brick walls, so it is easier for heat energy to escape. Double glazing can help to save energy.

A double-glazed window has two panes of glass. In between, there may be:

- air – which is a good insulator, much better than glass;
- a vacuum (empty space) – which does not conduct heat energy at all.

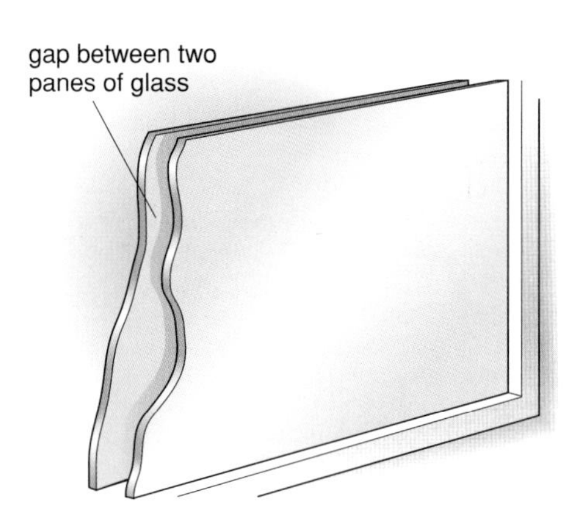

c **Which saves more energy, double glazing with air between the panes of glass, or with a vacuum between them?**

Questions

1 **List four ways heat energy can be conducted out of a house.**

2 **From the list, choose the material which is: a the best insulator, b the worst insulator.**

air vacuum glass

3 **Double-glazing can help a householder save money. From the list, choose *two* reasons why this is so.**

a **Heat energy is transferred more slowly through double glazing than through single-glazing.**

b **Double glazing is cheaper to install than single glazing.**

c **Heat energy cannot be conducted through the vacuum between two panes of glass.**

d **Heat energy cannot be conducted through glass.**

4 **Some disposable coffee cups are made from expanded polystyrene. Expanded polystyrene contains lots of air. Explain how the air:**

a **Helps the coffee stay hot for longer.**

b **Stops you from burning your fingers when you are holding a cup of hot coffee.**

Summary

- Heat may be transferred by conduction.
- Gases (such as air) are poor conductors of heat.
- Heat cannot be conducted through a vacuum.

9:2 Convection

It is a good idea to wear a hat on a cold day. Without a hat, heat energy from your head warms the air touching it. The warm air rises, carrying heat energy away. This is a **convection current**.

Wearing a hat will keep you warm. The hat insulates your head by preventing warm air from escaping.

Convection currents can also be useful. You may have an immersion heater at home; it warms the water in a hot-water tank. Convection currents make sure that all of the water gets hot. As warm water rises, colder water flows in to replace it.

Convection currents transfer energy through liquids and gases, but not through solids. The hot liquid or gas flows, carrying heat energy with it. A solid cannot flow.

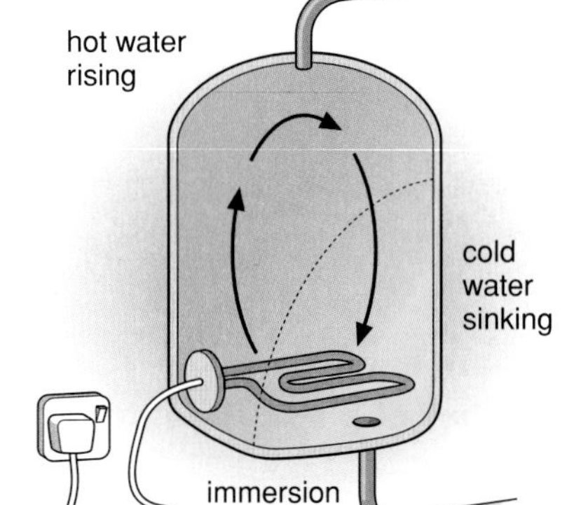

a Why must the immersion heater be at the bottom of the tank?

Convection in houses

It is not only us who can lose heat because of convection. Heat energy can escape from a house, carried by convection currents.

- A lot of the heat energy from a fire is carried up the chimney by rising air.
- Convection currents in the loft transfer energy to the roof tiles.
- There is air in the cavity of a wall. Convection currents in the air transfer heat energy from the inner wall, which is warm, to the outer wall, which is colder.

b If you stand on a table, your head is near the ceiling. The air up there feels warm. Use the idea of convection to explain why this is so.

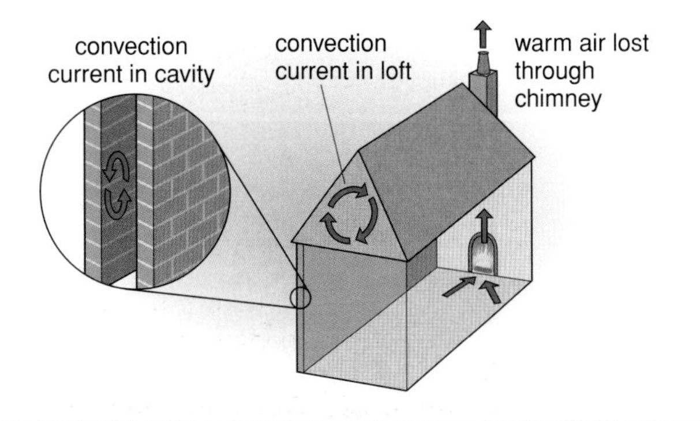

Did you know?
Eagles, vultures and hang-gliders use convection currents to fly high in the sky.

Cutting down on heat losses

Wasting heat energy will make our heating bills high, and it is not good for the environment. We can reduce the heat lost from our homes.

- Draught excluders can be fitted to doors and windows. These stop cold air coming into the house, and warm air escaping.
- Fitting loft insulation means that less heat energy conducts through into the loft space. Less energy is then transferred to the roof tiles.
- Rock wool or fibre glass fitted in the wall cavity prevents air from flowing around, so there are no convection currents to transfer energy from the house.

Loft insulation prevents heat escaping through the roof.

c Which of the houses in the photo have well-insulated lofts? How can you tell?

Questions

1 Copy the sentences below, choosing the correct words from each pair.

A convection current transfers energy from a warmer/cooler place to a warmer/cooler place.

Convection currents can flow in liquids and solids/gases.

2 Choose one of the following through which energy cannot be transferred by convection: air in a chimney; a glass window pane; air in a loft.

3 Draw a diagram to show how heat energy is transferred by a convection current in a loft. Label:

- warmer air rising;
- cooler air sinking.

4 Convection currents in cavity walls can transfer energy out of the house. How can this be prevented?

Summary

- Liquids and gases can carry heat by convection.
- Cavity wall insulation stops convection currents.

9:3 Radiation

If you sit in the sunshine, you feel heat energy coming from the Sun. How does this heat reach you through empty space? There is no air in space so it cannot be by conduction, because there is no material to conduct it. There is no liquid or gas in space so it cannot be by convection.

Heat from the Sun is carried by **radiation**. The hotter an object, the more energy it radiates. The Sun is very large and very hot, so it radiates a huge amount of heat energy.

This radiation is called **infra red radiation**. Because the Sun is very hot, it also radiates a lot of light energy.

A hot cup of tea is not hot enough to radiate light energy, but if you hold your hands close to the cup, you will feel the infra red radiation escaping from it.

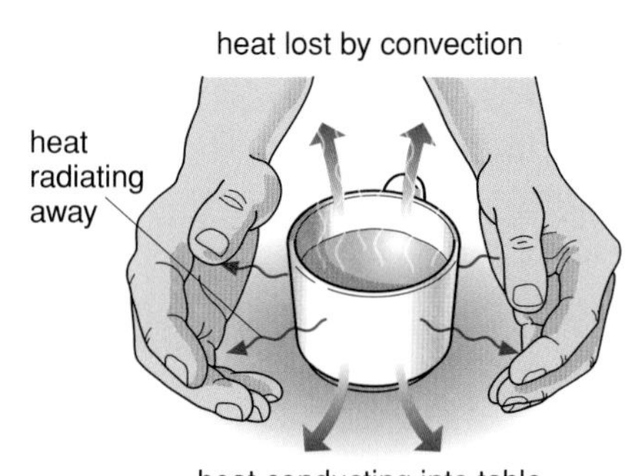

How heat escapes from a hot cup of tea.

a **List the three ways in which energy can escape from a hot cup of tea.**

Investigating radiation

Different colours radiate different amounts. This can be shown by an investigation. This metal cube has four differently coloured faces. Hot water is poured into the cube, and then the lid is put on. The temperature sensors detect infra red radiation coming from each face of the cube.

Each of the sensors S1, S2, S3 and S4 was connected to a computer which produced the four graphs shown on the right.

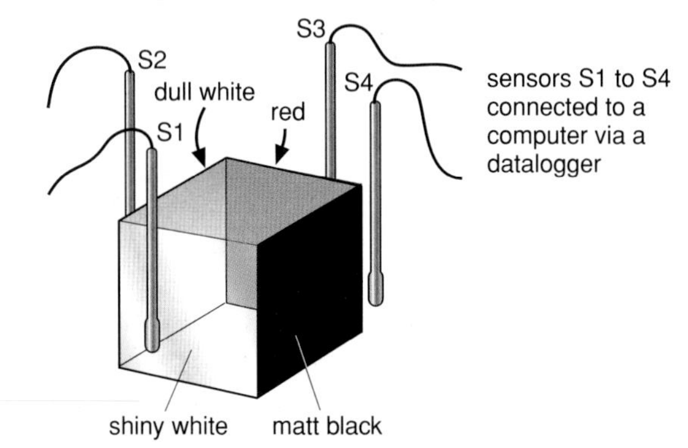

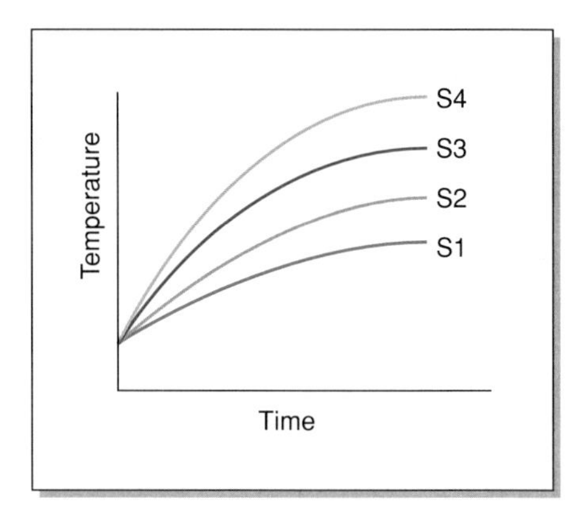

b **Look at the graphs on the right. Which sensor shows the fastest rise in temperature?**

c **Which colour gives out the most heat radiation? Which gives out the least?**

The texture of a surface will also affect the amount of heat radiated. Matt surfaces are better at radiating heat than shiny ones. The inside of an oven is usually black. When it gets hot, it radiates a lot of energy at the food which is cooking.

d **Should the inside of an oven be matt or shiny if it is to do its job well?**

Absorbing infra red

The colour of an object affects how well it takes in (absorbs) heat. These jars are filled with water and placed on the windowsill, in the sunshine. Heat from the sun will only warm the water inside if it is absorbed by the jar.

The temperature of the water in the matt black jar rises quickly. The temperature of the water in the shiny white jar rises more slowly.

This shows us that:

- Matt black surfaces are good absorbers of infra red radiation.
- Shiny white surfaces are poor absorbers of infra red radiation; instead, they reflect the radiation away.

e Is a matt black surface a good reflector of infra red radiation, or a poor reflector?

Saving energy by reflection

We can use what we know about absorbing and reflecting infra red radiation to make better use of energy in our homes.

Shiny metal foil is sometimes fixed to the wall behind a heater. This stops heat energy from being absorbed by the wall. It is reflected back into the room.

Sometimes shiny foil is fitted inside the loft, to reflect heat energy back into the house.

Did you know?

Snakes can see infra red radiation from their prey, so they can hunt in the dark.

Questions

1 **What is the name for the radiation emitted by hot bodies?**

2 **Copy the following sentences, choosing the correct word from each pair.**

A light, shiny surface is a good/poor absorber of radiation.

It is a good/poor reflector of radiation.

3 **The dashboard of a car is often made of matt black plastic. Why does it tend to get hot on a sunny day?**

4 **People who live in tropical countries often paint their houses white. Why do you think they do this?**

5 **Look at the experiment with the painted jars at the top of this page. What would you do to make sure that this experiment was a fair test?**

Summary

- Hot bodies emit infra red radiation. The hotter they are, the more they radiate.
- Good emitters: dark, matt surfaces.
- Poor emitters: light, shiny surfaces.
- Light, shiny surfaces are good reflectors (poor absorbers).
- Dark, matt surfaces are good absorbers (poor reflectors).

9:4 More for your money

If you could keep all the warmth in your house, you would save a lot of money. Energy escapes through the walls, floor and roof. It escapes every time you open the windows or the front door.

Amounts of energy are given in joules, J. The **joule** is the unit of energy.

Saving energy

The picture below shows a house which has been fitted with insulation. The figures tell you how many joules (J) of energy escape from the house each second on a cold day, before and after the insulation was fitted. How much energy is saved by installing cavity wall foam?

Energy lost each second before fitting insulation = 2500 J

Energy lost each second after fitting insulation = 1000 J

So, energy saved each second by fitting insulation = 1500 J

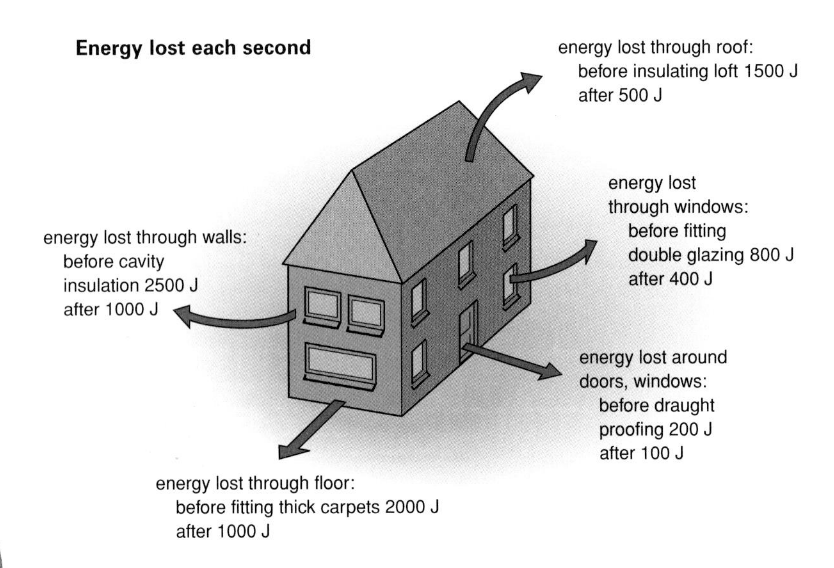

a Work out how much energy is saved by fitting insulation in the loft?

b Which saves more energy, cavity wall insulation or loft insulation?

Saving money

Most new houses are fitted with double glazing. This is good because it saves energy, stops draughts and cuts down on noise. However, it is expensive to fit double glazing to an old house, and it doesn't save much money. It is not very **cost-effective**.

If you live in an old house, it could be better to fit cavity wall insulation. Here are some figures to show how to decide whether cavity wall insulation is cost-effective:

Cost of cavity wall insulation = £600

Saving on heating bill in one year = £150

So, after 4 years, you will have saved the cost of the insulation. After that, you will save £150 every year. The **pay-back period** is 4 years.

Method of saving energy	Initial cost	Saving each year	Cost-effectiveness
cavity wall insulation	quite expensive	a lot	good value
loft insulation	quite cheap	a lot	very good value
thick carpets in all downstairs rooms	expensive	a lot	quite good value
draught proof doors	cheap	not much	very good value
double glaze windows	expensive	not much	poor value

c Which method of saving energy is the least cost-effective? Which are the most?

The initial cost of saving energy can be high. Eventually, you will save more money than you have spent. If you do not have much money, you may not be able to afford the initial cost. Then your energy bills will remain high.

Thinking of the future

Saving energy saves money. There are other reasons for using energy more efficiently.

- We don't want to use up all our reserves of fossil fuels too quickly.
- Burning fuels causes pollution which damages the environment.

Questions

1 Look at the pictures of the house on the opposite page.

a Before insulation is fitted, how much energy escapes each second?

b How much energy escapes each second when insulation is fitted?

c How much energy is saved by fitting insulation?

2 Give three reasons for fitting double glazing. Why are many people reluctant to fit it?

Summary

- Energy is measured in joules (J).
- It is important to save energy to reduce costs and to help the environment.

9:5 Making use of electrical energy

We use energy in different forms every day. Energy can help to make our lives easier or more comfortable.

If your house gets cold, you can use an electric heater to warm it up. The heater must be plugged into the mains electricity supply. Switch it on and it gets hot.

The heater is transferring energy to the room. It gets this energy from the electricity supply.

Electricity in the wires brings **electrical energy** to the heater; the heater transfers this energy as **heat energy**. (Another name for heat energy is **thermal energy**.)

This is an example of an **energy transfer.** We can show this transfer with an energy arrow diagram. The arrow here shows that all of the electrical energy which arrives at the heater is transferred as heat energy.

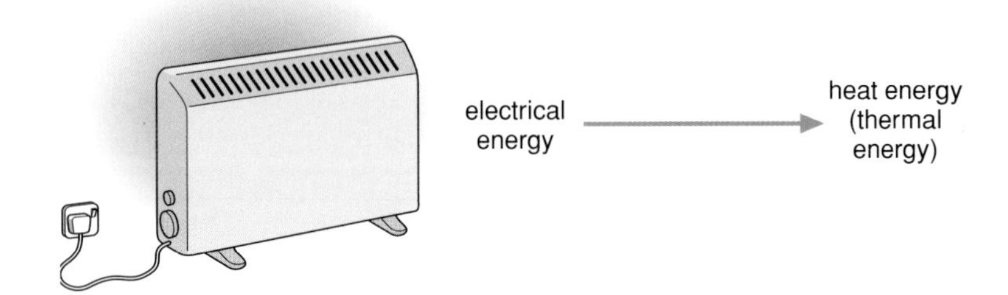

The heater transfers electrical energy as heat energy.

A cleaner car

The car in the photograph runs on electricity. It has rechargeable batteries to provide electricity for its motor.

The car is designed for use in cities. It doesn't produce exhaust gases as it travels around, so it helps to reduce pollution in city centres.

The car's electric motor makes the car move – it transfers electrical energy as movement energy. The scientific name for movement energy is **kinetic energy**.

The motor also gets hot. Some of the electrical energy is transferred as heat energy. This energy is wasted.

a The car's electric motor transfers energy in two forms, kinetic energy and heat energy. Which of these is a useful energy transfer?

Getting the picture

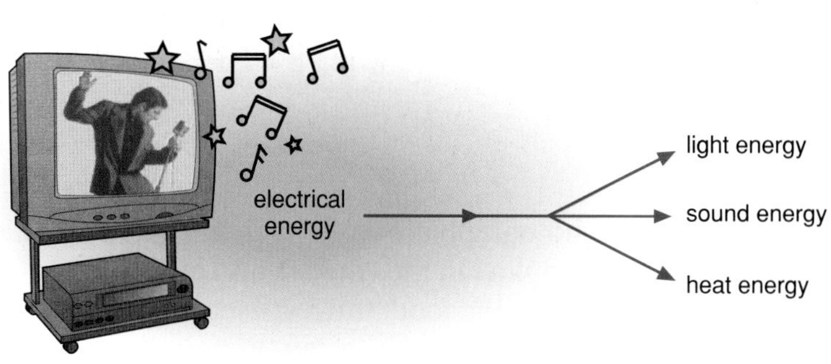

When you switch on a television set, you see a picture and hear sounds. The television is doing a useful job, providing pictures and sounds.

The television transfers electrical energy as two useful forms of energy: **light energy** and **sound energy**.

If you put your hand on the top of the television, you will find that it is warm. This shows that some of the electrical energy is transferred as heat energy. The energy arrow for the television set splits into three, to show the three forms of energy produced.

b **Draw an energy arrow to show the energy transfers for the electric car shown on the opposite page.**

Going to waste

Electrical devices such as heaters, motors and televisions are very useful. They can transfer energy at the flick of a switch.

Unfortunately, it is usually the case that some of the electrical energy transferred to them is wasted. It is transferred as a form which we do not want, usually heat energy. Some devices are noisy – they waste energy as sound. Sound energy is absorbed by the walls of the room, making them slightly warmer.

If you are watching the television, the waste heat energy which it produces makes the room warmer. Waste heat and sound energy both warm the surroundings.

c **A computer gets warm. It also makes a faint humming noise. What forms of waste energy are being produced?**

Questions

1 **kinetic electrical heat sound light thermal**

Copy and complete the sentences below, choosing words from the list to fill the gaps.

a **An electric heater transfers __________ energy as __________ energy.**

b **A radio transfers __________ energy as __________ energy.**

c **Another name for movement energy is __________ energy.**

d **Another name for heat energy is __________ energy.**

2 **An electric light transfers electrical energy as light and heat. Draw an arrow diagram to show this.**

3 **An electric mixer can be very noisy. What form of waste energy is it producing?**

4 **Look at the arrow diagram for a television set at the top of this page.**

a **What two forms of useful energy does it produce?**

b **What form of waste energy does it produce?**

Summary

- Electricity is a convenient way of transferring energy.
- Electrical devices can transfer electrical energy as:
  - heat (thermal energy);
  - light;
  - sound;
  - movement (kinetic energy).
- When energy is transferred, some is wasted because it is transferred in a form we do not want.
- This wasted energy makes the surroundings warmer.

9:6 Making better use of energy

The chefs in the photograph are working in a very hot kitchen. A lot of energy is supplied to the cookers and grills. But only a little is used in cooking the food; the rest escapes to the surroundings, making the kitchen a very hot place to work.

A microwave oven makes better use of the energy supplied to it. The electrical energy supplied to it is transferred as heat energy in the food. Only a little escapes; if you put your hand inside the oven after it has been used, you will find that it is not very warm.

This also means that a microwave oven cooks food more quickly than an electric or gas oven.

a Which wastes more energy, a microwave oven or a gas oven?

Efficient appliances

We want our electrical appliances to make good use of the electrical energy which is supplied to them. Electricity costs money, and we don't want to waste it.

A microwave oven transfers a lot of its energy to the food, more than a gas or electric oven. We say that the microwave oven is more **efficient** than the other ovens.

To decide how efficient a device (such as an oven) is, we need to think about two things:

- how much energy is supplied to the device;
- how much of this energy is transferred usefully.

An efficient device wastes less energy than an inefficient one.

The table compares two different light bulbs. One is much more efficient than the other.

		Amount of electrical energy supplied each second	Amount of energy transferred as light each second
	Ordinary filament lamp	100 J	10 J
	Low-energy lamp	20 J	10 J

From the fourth column of the table, you can see that both light bulbs produce the same amount of light energy each second.

From the third column, you can see that the filament lamp must be supplied with 100 J of energy each second. The low-energy lamp requires only 20 J each second.

b **The two light bulbs are of equal brightness. How can you tell this?**

c **Which light bulb is more expensive to run? Explain your answer.**

Calculating efficiency

The filament lamp wastes a lot of energy as heat energy. Only 10 J out of 100 J are transferred usefully, as light energy; 90% of the energy supplied to it is wasted.

The low energy lamp is more efficient, because it wastes less of the energy supplied to it. It transfers half of the energy supplied to it as useful light energy. We say that its efficiency is 50%.

The **efficiency** of a device tells us the fraction of the energy supplied to it which it transfers in a useful form.

d **Use the figures in the table to help you answer this. How efficient is the filament lamp: 1%, 10%, or 100%?**

Questions

1 **The unit of energy is written as J. What does this stand for?**

2 **A filament lamp is not very efficient. What happens to the energy it wastes?**

3 **A new car is more efficient than the old one it replaces. Does this mean that the new car wastes more energy than the old one, or that it wastes less energy?**

4 **An electric train wastes 30% of the energy supplied to it. A diesel train wastes 60%. Which is more efficient?**

Summary

- All devices waste some of the energy supplied to them.
- An efficient device wastes a smaller fraction of the energy supplied to it.
- The efficiency of a device is the fraction of the energy supplied to it which it transfers in a useful form.

9:7 The cost of electricity

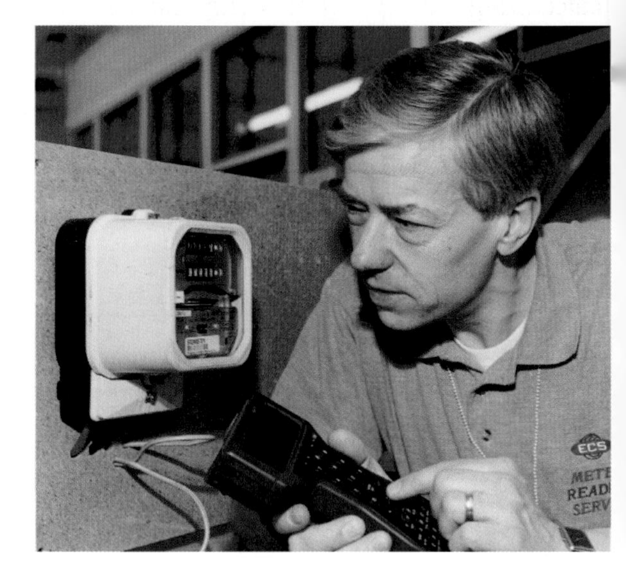

You have to pay for the electricity you use. The electricity meter records how many Units of electricity you have used.

The meter reader records the reading on the meter. By comparing this with the previous reading, it is easy to calculate the number of Units of electrical energy used.

Calculating the Units

To work out how many Units have been used, take the previous reading away from the new reading.

Present reading	= 30459
Previous reading	= 30329
Number of Units used =	130

Calculating the cost

The electricity bill shows the cost of each Unit. This is usually about 7p. Now you can work out the cost of the Units used.

Cost of 130 Units at 7p each = $130 \times 7p = 910p = £9.10$.

Here is the formula for calculating the cost of electricity:

Total cost = number of Units × cost per Unit

a What would be the cost of 100 Units if each Unit costs 7p?

Electrico plc

Ms C.U. Stomer
Ohm Sweet Ohm
Spark Street
Amperby

Meter reading	**30459**
Previous meter reading	**30329**
Units used	**00130**
Cost of each Unit	**7.0p**
Cost of Units used	**£9.10p**

Appliance power

Many electrical appliances are labelled with their electrical power. A heater might be labelled 1 kW or 2 kW. An electric cooker might be labelled 7 kW. The letters 'kW' stand for 'kilowatt'. This is the **power** of the appliance.

The power tells you how quickly the appliance transfers electrical energy.

A 1 kW heater transfers 1 Unit every hour.

A 2 kW heater transfers 2 Units every hour.

b How many Units of electrical energy does a 7 kW cooker transfer every hour?

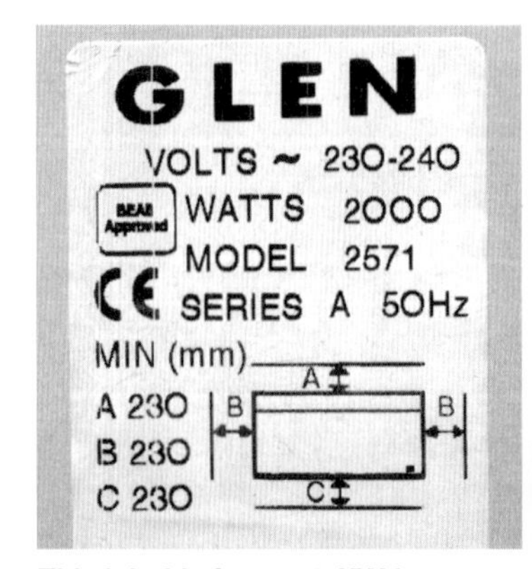

This label is from a 2 KW heater.

Kilowatt hours

The longer you use an appliance for, the more energy it transfers. If you use a heater with a high power for a long time, it will transfer a lot of energy. Here is the formula for calculating the amount of energy transferred:

Energy transferred (kWh) = power (kW) × time (h)

Because we are measuring power in kilowatts and time in hours, the energy comes out in kilowatt hours (kWh). A kilowatt hour is the same as a Unit.

1 Unit = 1 kWh

c How many kilowatt hours of electrical energy does a 7 kW cooker transfer in 2 h? How many Units is this?

> **Did you know?**
> Some electricity meters can be read remotely; they send their readings down the cables to the supply computer.

Questions

1 kWh kW Unit J

From the list:

a Choose *two* units which are the same as each other.

b Choose *three* units which are all units of energy.

2 A floodlight is labelled '2.5 kW'. How many Units of electrical energy does it transfer in one hour?

3 a A 3 kW electric kettle is used for 4 hours in a week. How many Units of electrical energy does it transfer in this time?

b If each Unit costs 7p, how much does this cost?

4 The pictures show the readings on the electricity meter in a holiday cottage, recorded at the beginning and end of a week's holiday.

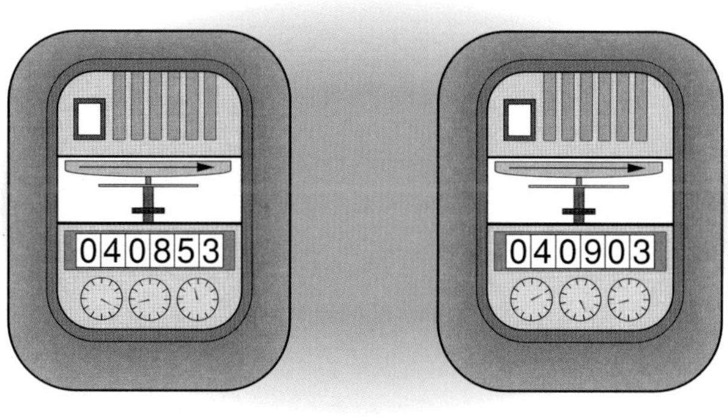

a What was the reading at the end of the week?

b How many Units were used during the week?

c If each Unit costs 7p, what was the total cost of the electricity used?

Summary

- The power of an electrical appliance is often given in kilowatts (kW).
- A 1 kW appliance transfers 1 Unit in 1 hour.
- 1 Unit = 1 kilowatt hour (kWh).
- Energy transferred (kWh) = power (kW) × time (h).
- Total cost = number of Units × cost per Unit.

9:8 More about power

The bedroom in the picture is a hot place to work. The heater transfers lots of energy to the room quickly. The other appliances transfer energy more slowly.

To make the room cooler, the girl could turn off the heater. Then the room would soon cool down.

A heater is labelled with its power in kilowatts (kW). The power of a light bulb is much less than a kilowatt. It is given in watts (W). Each bulb in the photograph is connected to the mains electricity supply. The brighter the bulb, the greater its power.

The 60 W bulb transfers 60 J of electrical energy every second. The 100 W bulb transfers 100 J of electrical energy every second.

a How many joules of energy are transferred by the other two bulbs every second?

Watts and kilowatts

A small torch bulb has a power of about 1 watt (W). This means that it transfers 1 J of energy every second. A watt is a joule per second.

A 1 kW electric heater has a much higher power than a torch bulb. A kilowatt is much more than a watt. A power of 1 kilowatt (kW) means that 1000 J of energy are being transferred every second.

So a kilowatt is a thousand watts: 1 kW = 1000 W

b How many watts make 2 kW?

Calculating power

The power of an appliance tells us how quickly it transfers energy. To find the power of an appliance, we need to know how many joules of energy it transfers every second.

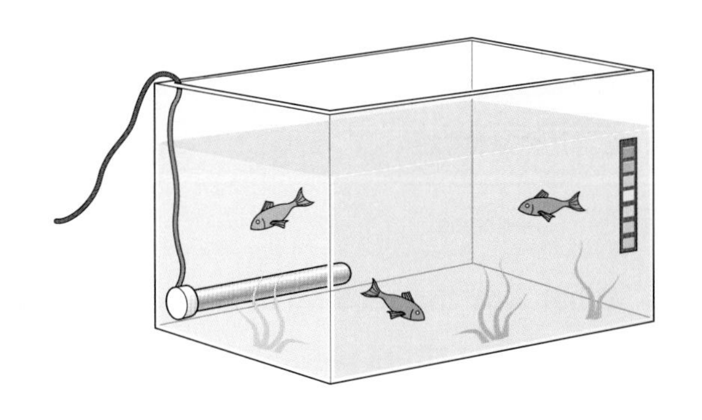

The fish-tank heater in the picture transfers 50 J of energy to the water in 10 s. What is its power?

We need to work out the number of joules transferred by the heater in 1 s:

Power of heater = 50 J/10 s = 5 W

The heater transfers 5 J of energy every second, so its power is 5 W.

Here is the formula for calculating the power of an appliance:

$$\text{Power (W)} = \text{energy transferred (J)}/\text{time (s)}$$

c **A more powerful heater transfers 100 J of energy in 10 s. How much energy does it transfer in 1 s? Now it is easy to answer: what is its power, in W?**

Calculating energy

When you switch on a light, it starts transferring energy to the room. The longer you leave it on for, the more energy it transfers.

The power of the light bulb tells you how much energy it transfers each second. The greater its power, the more energy it transfers.

A 100 W light bulb transfers 100 J each second. How much energy will it transfer in 20 s?

$$\text{Energy transferred in 20 s} = 100\text{ W} \times 20\text{ s} = 2000\text{ J}$$

Here is the formula for calculating the energy transferred by an appliance:

$$\text{Energy transferred (J)} = \text{power (W)} \times \text{time (s)}$$

d **A stereo system has a power of 50 W. How many joules of energy does it transfer each second? How much energy will it transfer in 10 s?**

Questions

1 **From the list, choose two quantities which are equal:**

100 W   1 kW   10 kW   1000 W

2 **Copy and complete the sentence below, which explains what we mean by 1 watt.**

If the power of an appliance is 1 watt (W), it transfers 1 ____________ of energy in 1 ____________.

3 **John uses a 2 kW electric heater to heat his flat. He is worried about his electricity bills. Which of the following would help to reduce the amount of electricity he uses?**

a **Using a 3 kW heater.**

b **Using a 1 kW heater.**

c **Switching the heater on for less time.**

Summary

- 1 watt (W) is 1 J of energy transferred in 1 s.
- A kilowatt is a thousand watts (1 kW = 1000 W).
- Power (W) = Energy transferred (J)/time (s).
- The amount of electrical energy transferred by an appliance depends on both its power and how long it is switched on for.

Energy transferred (J) = power (W) × time (s)

9:9 Going up

The skiers on the chair-lift are taking the easy way to the top of the mountain. Large electric motors pull the line of chairs up the steep slope. When they reach the top, the skiers can enjoy a fast run back to the bottom of the ski slope.

Gaining energy

The chair-lift transfers energy to the skiers. The higher they are lifted, the more energy has been transferred to them. The skiers at the top of the run have a lot of stored energy. We know they have this energy because, when they push themselves off, they start going faster and faster.

At the top of the slope, the skiers have a store of **gravitational potential energy**. (We can write this as **GPE**, for short.)

Their store of GPE gets less and less as they speed downhill.

a Where do the skiers have most GPE?

b Where do they have least GPE?

Lifting a load

This tall crane is lifting a load of bricks to the top of a tall building. First, it lifts them straight up. It is transferring energy to them, and their GPE is increasing.

Then the crane swings the bricks to the side, to reach the top of the building. The bricks are not getting higher, and their GPE is not increasing.

There is friction in the movement of the crane. Some of the energy it transfers is wasted as heat energy. The energy arrow shows this.

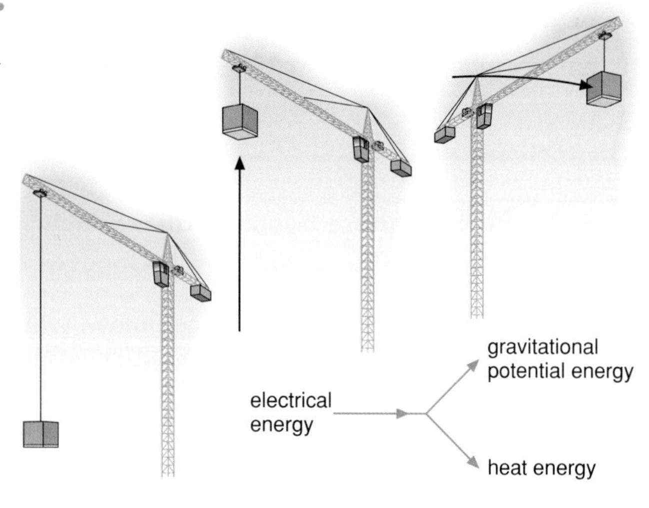

c The crane has an electric motor. It transfers electrical energy. In what two forms is this energy transferred?

Sliding downhill

The children waiting at the top of the flume have a store of GPE. When they reach the bottom of the slide, they are moving fast. Some of their GPE has been transferred as movement energy (kinetic energy). Some of their GPE is wasted as heat energy, because there is friction between the child and the slide.

d Draw an energy arrow to show the energy transfers as a child slides down the flume.

Turning the millwheel

In the past, many factories and mills relied on water stored behind a dam to provide the energy needed to operate machinery. The water in the millpond stores GPE. When it is released, it turns a waterwheel. The wheel turns the machinery in the mill.

Today, we store water behind high dams to generate electricity. The water flows down through a turbine which makes a generator spin. This is a hydroelectric scheme.

This watermill makes use of energy stored by water. Water in the millpond is a store of gravitational potential energy. As it flows downhill, the moving water turns the mill machinery.

Questions

1 What do the letters GPE stand for?

2 Copy the sentence below, choosing the correct word from the pair:

As an apple falls to the ground, its GPE increases/decreases.

3 Look at the picture of the diver.

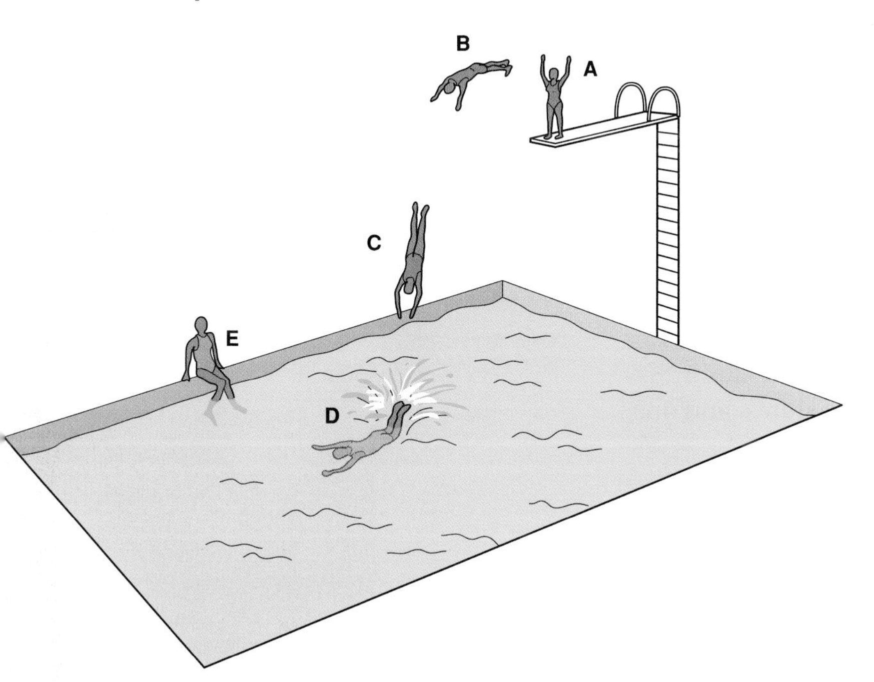

a At what point does she have the most GPE?

b At what point does she have the least GPE?

4 A racing skier wants to go as fast as possible. What *useful* energy transfer happens as a skier comes downhill?

Summary

- An object stores gravitational potential energy (GPE) when it is lifted up.
- The higher it is lifted, the greater its increase in GPE.

9:10 Fossil fuels to electricity

Every time you use a mains-powered electrical device at home, energy is transferred to your home from a power station.

This giant power station produces over a million Units of electricity every hour. That's enough electricity for quite a big city. In the picture you can see:

- the store of coal, waiting to be burned;
- the chimneys, which release waste gases from the burning coal;
- the cooling towers from which waste heat escapes.

Inside the power station

Burning coal is used to boil water. The steam produced by the **boiler** is very hot – over 100° C. It rushes along pipes to the turbines.

The moving steam causes the **turbines** to spin rapidly, like wind blowing on a windmill.

The turbines make the **generators** spin. Thick cables (wires) carry the electricity away from the generators.

a Which machine produces the electricity in a power station?

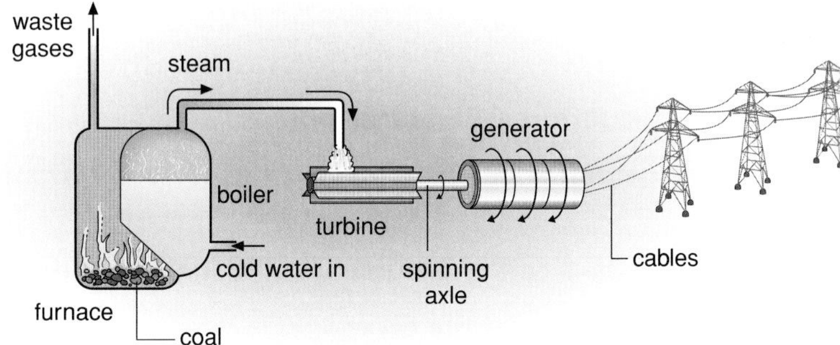

How electricity is generated – a coal-fired power station.

Fossil fuels

Electricity has helped to make our lives easier, but it's not all good news.

Most power stations in the UK burn coal, oil or gas. These are fossil fuels.

Fossil fuels are formed from the remains of plants and animals which died millions of years ago. They are stores of ancient chemical energy. Once they have been burned, we cannot replace them. They are **non-renewable** energy stores, and one day we may run out of them.

The greenhouse effect

The atmosphere is like a blanket around the Earth, keeping it warm. This is the **greenhouse effect**.

When fossil fuels are burned, carbon dioxide gas (CO_2) is released into the atmosphere. Coal produces the most carbon dioxide.

Extra carbon dioxide in the atmosphere is a problem. It traps

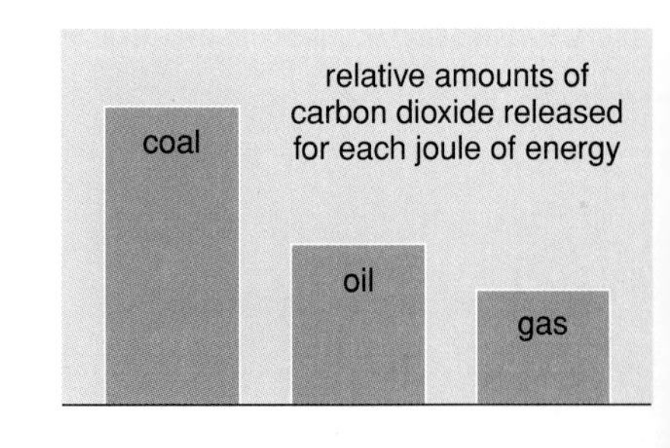

more heat, increasing the greenhouse effect and making the Earth hotter. There is worrying evidence that the Earth's average temperature rose in the last 100 years, and is rising faster today. Most scientists believe our climate is being affected. It may also make sea levels rise, swamping some low-lying parts of the world.

b **Which produces the least carbon dioxide, oil or gas?**

Acid rain

Coal and oil often contain sulphur. When this burns, it becomes sulphur dioxide (SO_2). This is released into the atmosphere through the power station's chimney.

Sulphur dioxide is acidic. When it dissolves in rain, the result is **acid rain**, which pollutes the environment. Plants die; fish and other water creatures are poisoned.

There are two ways to avoid this:

- remove the sulphur before burning the fuel;
- remove the sulphur dioxide from the power station's waste gases.

Both ways are expensive, and add to the cost of using coal and oil.

These trees have died because acid rain has poisoned the ground in which they are growing.

c **What substance is produced when sulphur burns? Is it acidic or alkaline?**

Questions

1 **Name three fossil fuels.**

2 **Copy and complete the diagram, which shows how a fossil fuel power station works.**

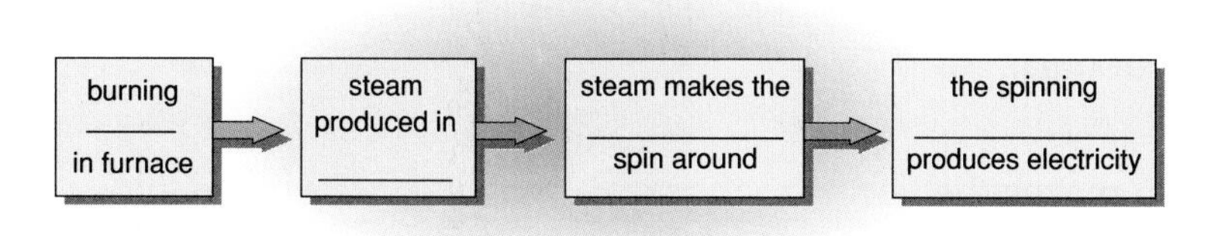

3 **When fossil fuels are burned, polluting gases are produced. Which gas:**

a **increases the greenhouse effect?**

b **contributes to acid rain?**

4 **Put these fuels in order, starting with the one which produces most carbon dioxide: coal, gas, oil.**

Summary

- Coal, oil and gas are fossil fuels.
- When fossil fuels are burned, gases are released which can pollute the environment.
- Carbon dioxide increases the greenhouse effect and adds to global warming.
- Sulphur dioxide causes acid rain.
- Coal produces the most carbon dioxide; natural gas produces the least.

9:11 Nuclear power stations

As supplies of fossil fuels run out, it becomes necessary to look for other stores of energy.

A nuclear power station contains a reactor which uses uranium or plutonium as its fuel. It does not burn fossil fuels.

Uranium and plutonium are radioactive substances. They gradually decay, releasing energy. In a nuclear reactor, their decay is speeded up so that they release energy much more quickly.

The photograph shows the giant machine which removes used fuel from the reactor, and puts new fuel in its place.

Inside the power station

As uranium decays in the reactor, it releases lots of energy. This is used to boil water, just like in a coal-fired power station.

a Which part of the nuclear power station contains the uranium fuel?

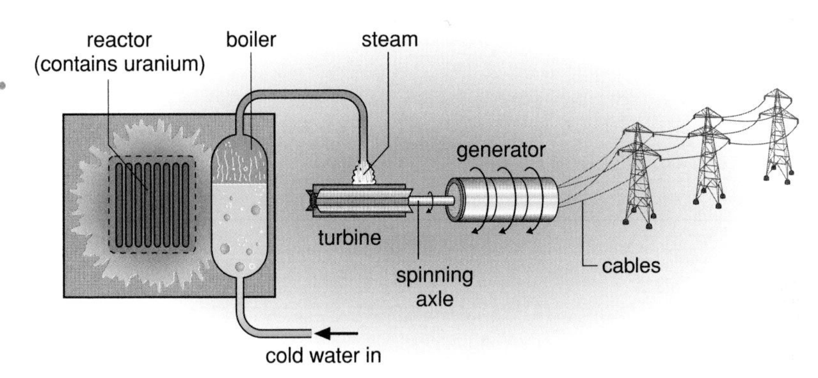

Radioactive decay

In a nuclear power station, the uranium fuel does not burn. No carbon dioxide is produced.

As the uranium decays, it produces radiation, which makes the uranium hot. It also produces radioactive waste products, which are very hazardous.

In normal operation, very little radiation or radioactive material escapes from the power station. However, if there is a serious accident, large amounts of radioactive material may escape. It can be spread over large areas.

b Which of the following are produced when a nuclear fuel such as uranium decays: radiation, carbon dioxide, radioactive substances, sulphur dioxide?

This nuclear power station, at Chernobyl in the Ukraine, exploded in 1987. Radioactive material spread over several villages and large areas of farmland. Many hundreds of deaths resulted.

Radioactive waste

The waste products are hazardous because of the radiation they produce. They keep on decaying for thousands of years.

There are two ways of storing radioactive waste:

- bury the waste deep underground;
- store it in steel containers at ground level, where it can be carefully monitored.

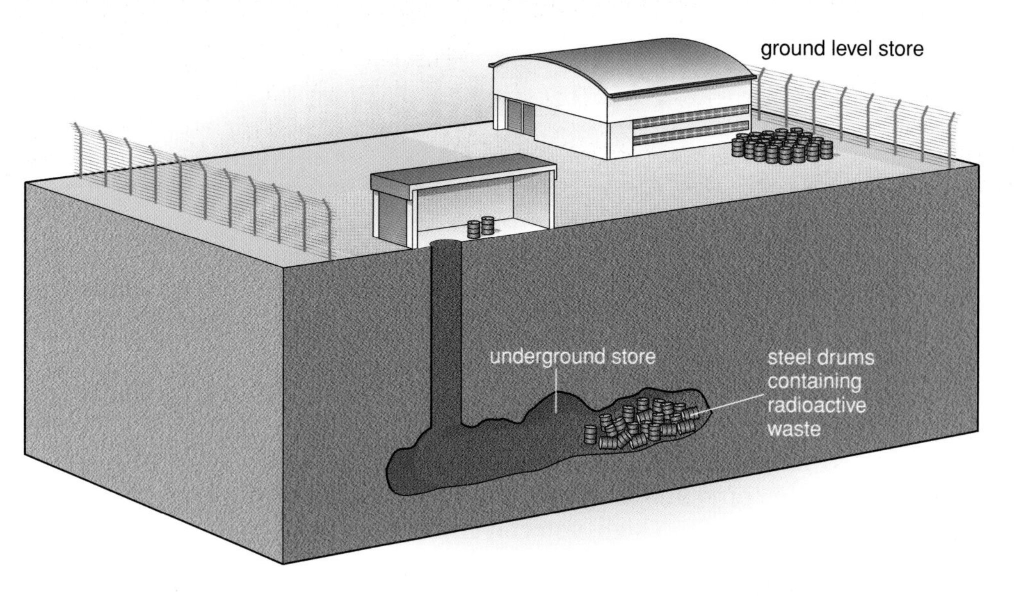

It is difficult to be sure which is the safest way.

The radiation from radioactive substances can cause cancer. The more of this radiation someone is exposed to, the greater their chance of contracting cancer.

Uranium comes from mines. Eventually, we may use up all the available stores of uranium. It is a non-renewable fuel.

c Why must waste from a nuclear power station be stored carefully for a very long time?

Did you know?
Nuclear power adds only 0.1% to our exposure to dangerous radiation.

Questions

1 **Name two fuels used in nuclear power stations.**

2 **The following sentences explain how electricity is generated in a nuclear power station. Copy them out, choosing the correct word from the list to fill each gap.**

generator boiler reactor turbine

Uranium fuel in the __________ releases a lot of energy.

This energy heats water to produce steam in the __________.

The steam makes the __________ spin around.

This turns the __________ which produces electricity.

3 a **Which gas produced by coal-fired power stations increases the greenhouse effect?**

b **Does a nuclear power station produce this gas?**

Summary

- Nuclear power stations use uranium or plutonium as their fuel.
- Nuclear fuels do not produce the same polluting gases as fossil fuels.
- In an accident, large amounts of radioactive material may be spread over a wide area.
- The waste produced by a nuclear power station can stay dangerously radioactive for thousands of years, and must be stored safely.

9:12 Geothermal and wind energy

Fossil fuels and nuclear fuels are non-renewable. We must look for other sources of energy to supply our demand.

This warm mudbath is a tourist attraction at Arboletes in Colombia, South America. Heat comes up from radioactive rocks deep underground.

Heat from the Earth

The holidaymakers in the photograph are enjoying a hot mudbath. In some places, particularly near volcanoes, hot water and steam come bubbling up from inside the Earth. If the temperature is not too high, you can enjoy its warmth.

Deep underground, radioactive substances such as uranium in the rocks are decaying. They release their energy, and this makes the rocks hot. Water in cracks in the rocks boils.

Geothermal power

A **geothermal power station** makes use of this energy. The steam turns a turbine and generator. For this to work, the steam must be at a high temperature. If the temperature is low, the hot water may be used for heating people's homes.

a Hot springs produce steam for thousands of years. Is this a renewable or non-renewable energy source?

Iceland has many hot springs (and volcanoes). This geothermal power station makes use of this natural source of steam.

Moving air

The wind is moving air. It has kinetic energy. This energy can be captured by a **wind generator**.

The large blades of the turbine spin in the wind. The turbine makes a generator turn, to produce electricity.

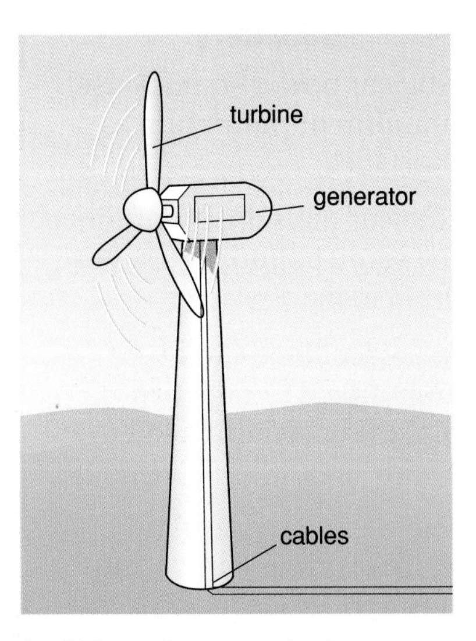

b What does a wind generator generate?

Did you know?
A farm in Yorkshire uses wind turbines to power its milking parlour and its yoghurt-making machines.

Wind farms

Wind generators are built in windy places:

- on hillsides;
- on the coast.

Several generators together make up a **wind farm**.

This wind farm is in the Pennine Hills of northern England.

If you are close to a generator, you may hear its blades turning around. If you live nearby, you may consider this a form of noise pollution.

To many people, a wind farm may spoil the look of the landscape – a form of visual pollution.

c Which part of a wind generator is the turbine?

Questions

1 Copy the sentence below, using words from the list to fill the gaps.

radioactive geothermal heat

A __________ power station makes use of __________ energy released by the decay of __________ substances inside the Earth.

2 Give *two* ways in which a wind farm may damage the environment.

3 Why is the coast a good place to build a wind farm?

4 Choose words from the list to answer the questions which follow.

reactor generator cables turbine

a Which part of a wind generator is made to spin round by the wind?

b Which part produces electricity?

Summary

- Steam and hot water from inside the Earth can be used to heat water in geothermal power stations.
- This energy comes from the decay of radioactive elements such as uranium, inside the Earth.
- The energy of moving air (the wind) can also be used to generate electricity.
- Wind farms are sited in windy places, such as on the coast or on hills.
- Wind farms create noise pollution. Some people think they are ugly.

9:13 Energy from water

The giant dam in the photograph is storing water for a **hydroelectric scheme**. The water behind the dam stores a lot of gravitational potential energy (GPE). When the water flows through the hydroelectric station, it generates a lot of electricity.

The Hoover Dam on the Colorado River, USA, generates more than 1 million kW of electricity.

Turning turbines

The water flows past the blades of a turbine. The water's GPE is transferred to the turbine as kinetic energy.

As in other power stations, the turbine turns a generator, which produces electricity.

a The energy arrow shows the energy transfers in a hydroelectric power station. Which form of energy is the scheme designed to produce? Which two forms are wasted energy?

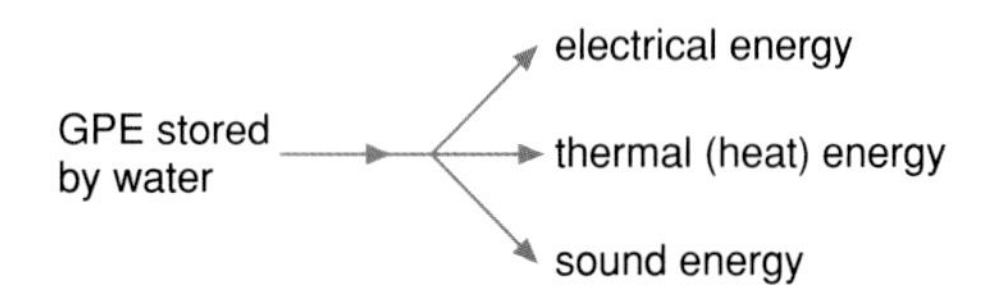

Impact on the environment

A hydroelectric scheme is usually built high up, in a hilly area. A dam is built across a river to create a reservoir.

When a reservoir is created, it floods a large area of land. In the past, the land might have been used for farming, or for forestry.

People who lived or worked on the land have to move away. Wildlife may be harmed. This can lead to protests when a new hydroelectric scheme is proposed.

b How might a farmer suffer when a new hydroelectric scheme is built?

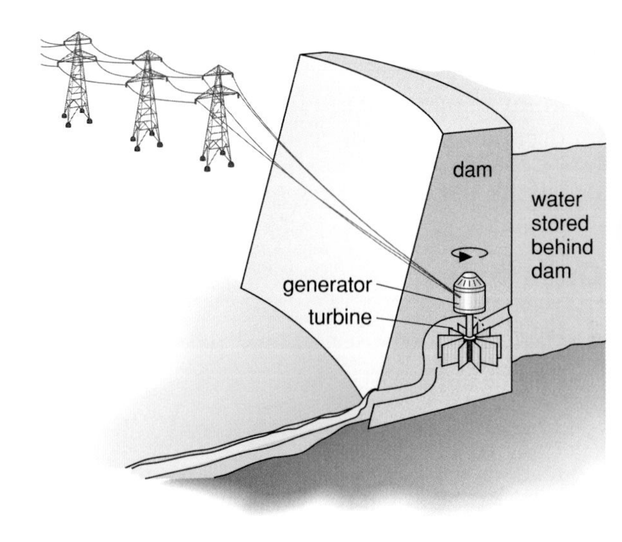

Energy from the tides

A river estuary is a good place to build a tidal power scheme. When the tide comes in, the estuary fills with water. A barrage (a dam) can be built across the estuary to trap the water.

When the tide goes down again, the water can be released through a tidal power station. The water turns turbines which turn generators.

A river estuary is also a good place for wading birds. There are mudflats which are the habitat of worms and other creatures. As the tide goes out, the birds can feed on these creatures.

When a tidal barrage is built, the mudflats remain underwater. They are no longer a good source of food for the birds, which must move away. This means that the number of wading birds will decline.

c What form of energy is stored by water trapped behind a tidal barrage?

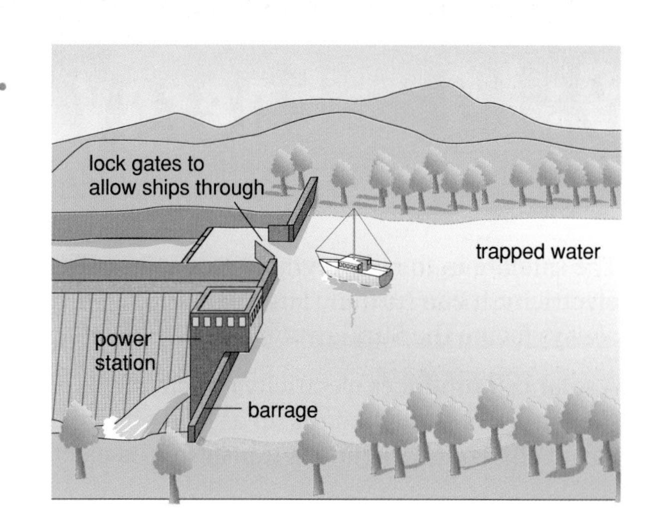

Energy from waves

Waves are moving water. In a wave generator, as the waves go up and down, they make a turbine spin. The turbine drives a generator.

Did you know?

Questions

1 The boxes show the energy transfers in a tidal power scheme. Copy these, putting them in the correct order.

electrical energy (carried by electricity in cables)	GPE (of water stored behind barrage)	kinetic energy (of spinning generator)	kinetic energy (of spinning turbine)

2 Here are some people who might be affected when a new hydroelectric scheme is built:

farmer angler (fisherman) forestry worker windsurfer

a Who might be harmed by the new scheme?

b Who might benefit from the new scheme?

3 Which one of the following is a good reason for building a tidal power station?

a It makes use of a non-renewable energy resource.

b We can use less fossil fuels.

c A large area of land is flooded.

Summary

- Electricity may be generated by allowing water to flow downhill through a turbine, which turns a generator; or by using waves to drive a generator.
- In a hydroelectric scheme, water from an upland river is trapped behind a dam.
- In a tidal power scheme, water is stored behind a barrage built across a river estuary.

9:14 Energy from sunlight

The satellite in the photograph uses solar cells as its source of electricity. It can turn the large panels of cells so that they are always facing the Sun.

A solar cell produces electricity directly from sunlight. Most of the energy of sunlight ends up as heat energy (it warms the solar cells), but a fraction is transferred as electrical energy.

a **A particular solar cell transfers one-fifth of the light energy which falls on it as electrical energy. What is its efficiency? Choose from 0.5%, 2%, 5%, 20%, 50%.**

On a sunny day

We can also use solar cells on Earth. They work best in bright sunlight. At night they produce no electricity. The brighter the light falling on them, the more electricity they produce.

Some garden fountains use solar cells to provide the electricity which operates them.

On a sunny day, the fountain starts to work. On a dull day, the fountain stops working.

b **Why do the fountains stop working at night?**

Expensive electricity

Sunlight, the 'fuel' for solar cells, is free. However, at present, solar cells are expensive. This means that, over the lifetime of a solar cell, the cost of each Unit of electricity it produces is high.

People have to think carefully about where to use solar cells. They are useful in remote places, where it is difficult to supply electricity by any other means.

Space is a remote place. That is why many satellites use solar cells. In remote parts of Australia, roadside phones are powered by solar cells. This is much cheaper than laying cables for hundreds of kilometres.

Solar cells are also useful where only a little electricity is needed. That is why some watches and calculators use solar cells.

A solar-powered phone box in Australia.

Comparing costs

The diagram shows the cost of electricity from some different sources. You can see that solar cells provide expensive electricity.

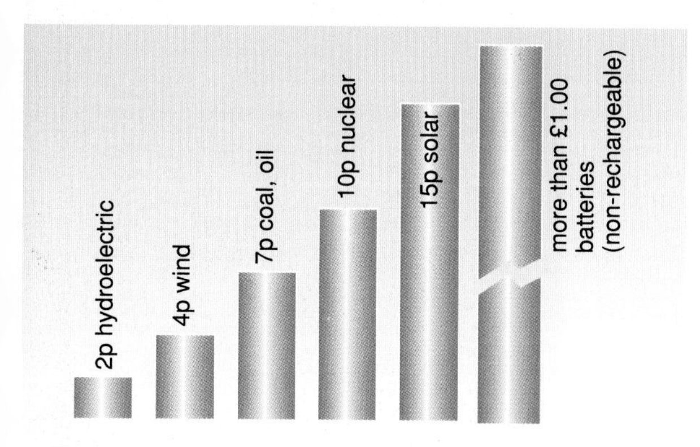

c Which source of electricity is more expensive than solar cells?

Questions

1 Copy and label this energy arrow diagram which shows how a solar cell transfers energy.

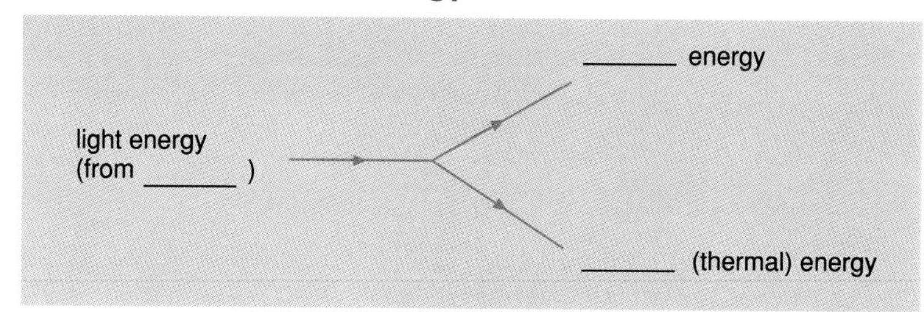

2 Copy the graph which shows best how the amount of electricity produced by a solar cell depends on the brightness of the light. Explain why you have chosen this graph.

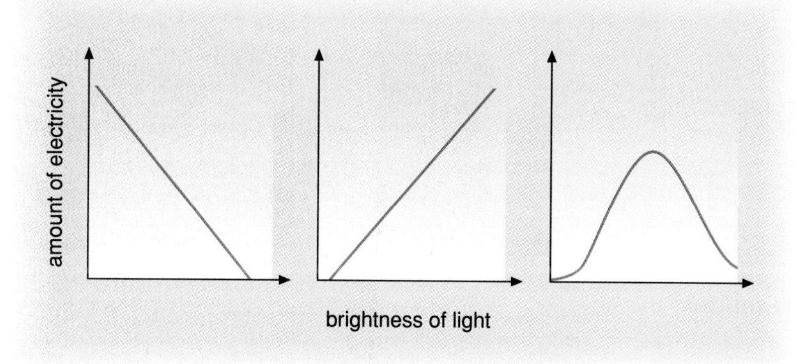

3 An emergency telephone is sited on a high mountain pass, for use by walkers when they are in trouble. Why would solar cells be a good choice to provide the electricity for this phone?

Did you know?
In the Solar Challenge race, solar-powered cars race over two thousand miles across Australia, from Darwin to Adelaide.

Summary

- Solar cells produce electricity directly from sunlight.
- Solar cells are expensive; this means that the electricity they produce is expensive.
- Solar cells are useful in remote places (e.g. on satellites) and in watches and calculators.

9:15 Electricity when you need it

A nuclear power station runs all day and all night, generating electricity. It is very **reliable**, because we can be sure that it will keep producing electricity whatever the weather.

Other power stations which use fuel (coal, oil and gas) are also reliable.

Electricity from renewable sources may depend on the weather:

- solar cells only work in daylight, and they work best in bright sunlight;
- wind generators only work when the wind is blowing;
- hydroelectric schemes may run out of water when there is a drought.

This means that these sources of electricity are **unreliable**.

Tidal power stations depend on the tide, so they produce more power at some times than at others.

a Which one of these sources of electricity depends on the weather?

coal-fired power station **tidal power station** **wind farm**

When the wind blows, the small wind generator supplies electricity which is used to charge up a battery.

Demand for electricity

We use more electricity at some times of the day than at others:

- There is high demand for electricity in the morning, when people are getting up and having breakfast, and in the evening when they are cooking. People also use a lot of electricity when they are at work.
- There is low demand at night, when people are asleep.

We also use more electricity in the winter, when it is cold, than in the summer.

Electricity companies need to start up extra power stations in the morning. The diagram on the right shows that nuclear power stations take the longest time to start up.

nuclear	– longest start-up time
coal	
oil	
gas	– shortest start-up time

For this reason, electricity companies keep nuclear power stations running all the time. It is easier to switch the gas-fired stations on during the day and off at night.

b Which type of power station can be started up most quickly?

c At what time of day do you think most electricity is used by homes?

High demand

Sometimes, a lot of electricity must be generated very quickly. Here is how the electricity companies meet this sudden demand:

- At night, they use spare electricity to pump water uphill into a reservoir.
- Then, when demand is high, they release the water so that it flows downhill and generates electricity.

This kind of hydroelectric scheme is called a **pumped storage scheme**. This avoids surplus (spare) electricity being wasted at night.

d Why is there surplus electricity at night?

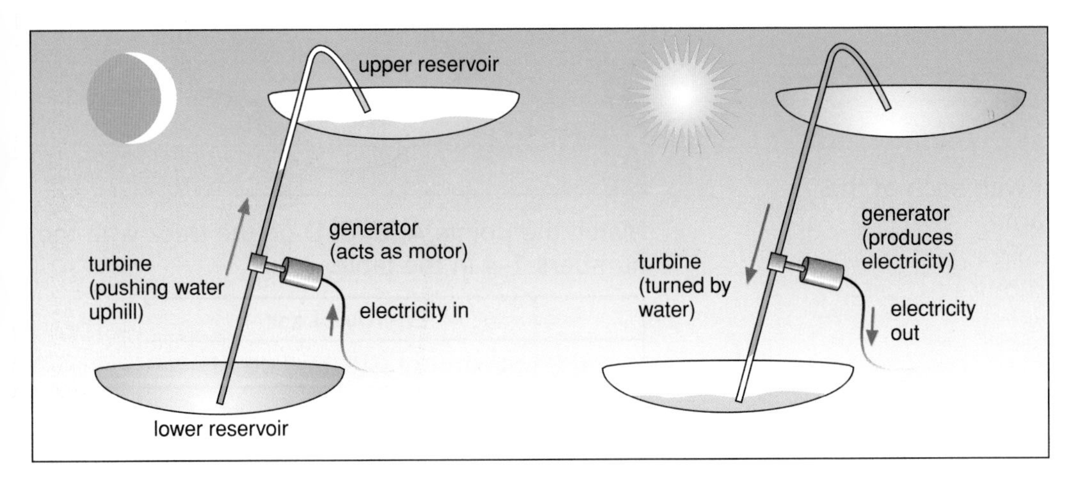

Questions

1 **Some power stations can be started up more quickly than others. Put these power stations in order, starting with the one with the longest start-up time.**

coal-fired oil-fired nuclear gas-fired

2 **Which of these sources of electricity is the most reliable?**

coal-fired power station tidal power station wind farm

3 **A solar cell may be described as an *unreliable source* of electricity. Choose the correct reason for this from the list:**

a **Solar cells are expensive to make.**

b **Solar cells produce electricity as soon as sunlight falls on them.**

c **Solar cells only produce electricity when sunlight falls on them.**

4 **In a pumped storage scheme, energy is stored by pumping water into a high reservoir. Which type of energy has increased when the water has been pumped uphill?**

a **The kinetic energy of the water.**

b **The electrical energy stored by the water.**

c **The gravitational potential energy stored by the water.**

Summary

- Power stations which use fuels are very reliable, because they can run steadily for weeks on end.
- Nuclear power stations are the slowest to start up.
- Hydroelectric schemes can work in reverse, using spare electricity to pump water uphill.
- The amount of electricity generated from other sources may depend on the strength of the wind, the brightness of sunlight and the variations of the tide. These sources are unreliable.

End of module questions

1 The diagram shows the main parts of a nuclear power station.

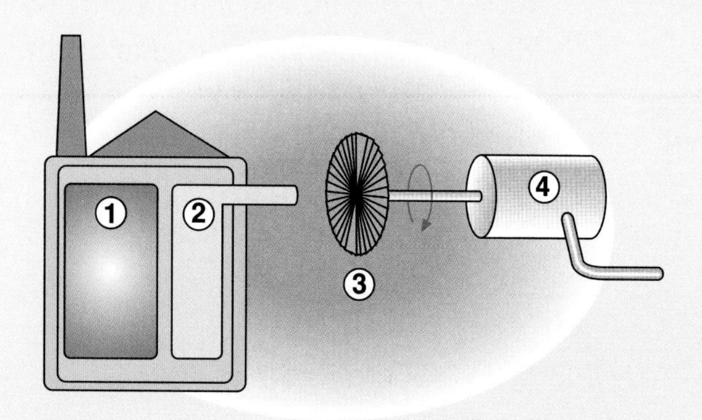

Match words from the list with each of the numbers 1–4 in the diagram.

generator
reactor
turbine
boiler

2 The sentences below describe some of the different sources of energy we use.

Choose words from the list for each of the spaces 1–4 in the sentences.

economical
fossil
non-renewable
renewable

The energy of the wind will always be available; it is a __1__ energy source.

Most of the energy we use comes from __2__ fuels.

When they have been burned, they cannot be replaced; they are __3__ energy sources.

The more __4__ we are with them, the longer they will last.

3 We use different devices to transfer electrical energy to other useful forms of energy.

Match devices from the list with each of the numbers 1–4 in the table.

grill element
light bulb
loudspeaker
motor

1	sound
2	movement
3	light
4	heat

4 When a fairground car runs down the track, energy is transferred between gravitational potential energy (GPE) and kinetic energy (KE).

The diagram shows the track that the car follows. It stops at point D.

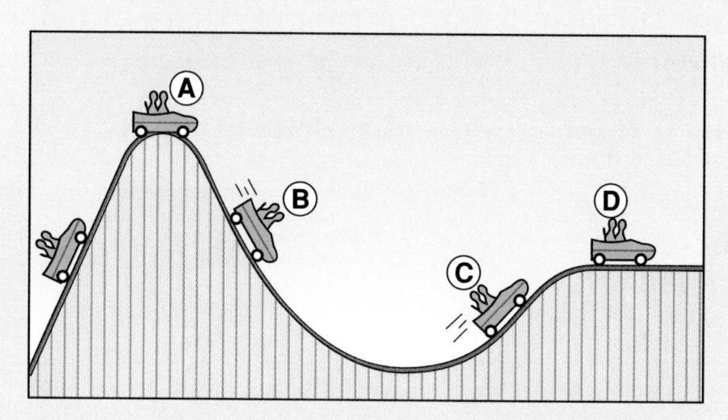

Match the points A, B, C, D on the track with the numbers 1–4 in the table.

	Energy of car
1	maximum GPE and no KE
2	GPE but no KE
3	GPE decreasing, KE increasing
4	GPE increasing, KE decreasing

5 A 2 kW electrical heater is run for 5 hours. Each Unit of electricity costs 7p.

Match quantities from the list with each of the numbers 1–4 in the table.

7
10
70
2000

1	power of the heater (in W)
2	cost per kilowatt-hour (in p)
3	number of Units used
4	total cost of using heater (in p)

6 Electricity is generated in several different ways.

Which **two** of the following ways make use of the energy released by uranium as it decays?

solar cells
energy from hot rocks
tidal power station
coal-fired power station
nuclear power station

7 Most of the energy supplied to a television set is wasted.

Which **two** of the following forms of energy are usefully transferred by a television set?

electrical energy
heat
light
movement
sound

8 The diagram shows how electrical energy can be generated in a power station.

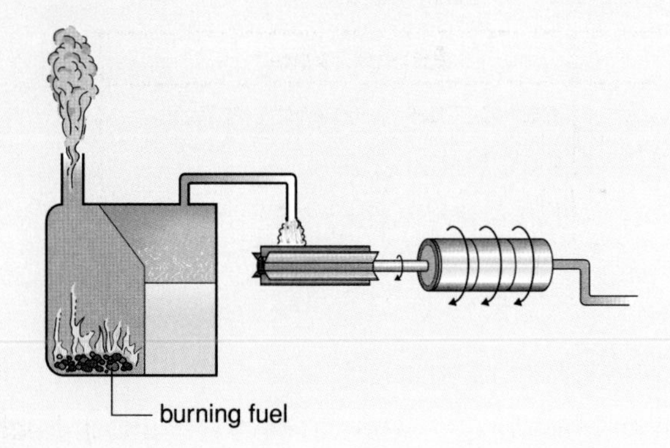

8.1 What is the main form of energy released when the fuel is burned?

A gravitational
B heat
C light
D movement

8.2 What is the main form of energy produced by the turbine?

A gravitational
B heat
C light
D movement

8.3 Which one of the following is a renewable fuel?

A coal
B gas
C oil
D wood

8.4 Electricity can also be generated using the flow of water from behind a tidal barrage.

What is the main form of energy which water has when it is stored behind a barrage?

A gravitational potential energy
B heat (thermal energy)
C light
D movement

9 Heat escapes from a warm house on a cold day.

9.1 In which way does heat leave the house through the floor?

A radiation
B conduction
C convection
D insulation

9.2 Air moves around in the loft, carrying away heat. In which way is heat being transferred?

A radiation
B conduction
C convection
D insulation

9.3 Insulating material made from glass fibres can be put into wall cavities to reduce heat loss. Why does this work?

A The insulated walls will be less damp.
B Glass is a better insulator than air.
C Air cannot flow in the wall cavity.
D Glass fibres are a good absorber of heat.

9.4 Blinds can be fitted to the windows to reflect away the Sun's rays on a hot day. Which of the following blinds would be best for this?

A shiny black blinds
B matt black blinds
C shiny white blinds
D matt white blinds

10 The diagram shows how much energy is transferred by a light bulb every second.

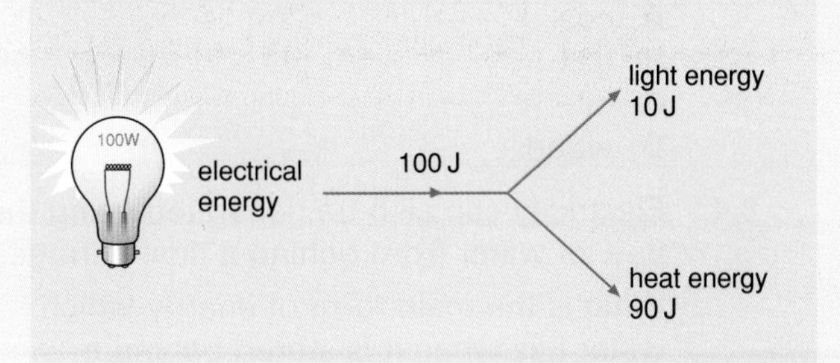

10.1 How much electrical energy is supplied to the light bulb?

A 1 J
B 10 J
C 89 J
D 100 J

10.2 How much energy is wasted by the light bulb?

A 1 J
B 11 J
C 89 J
D 90 J

10.3 What is the light bulb's efficiency?

A 1%
B 10%
C 90%
D 100%

10.4 What happens to the waste energy?

A It is absorbed by the light bulb.
B It remains in the electric current.
C It warms up the surroundings.
D It disappears.

Module 10 – Electricity

One of the first people to study electricity in a scientific way was Alessandro Volta. He was an Italian, working in the late eighteenth century. The volt is named after him.

At the time, meters such as ammeters and voltmeters had not been invented. When he generated static electricity, Volta would connect it to a metal bucket. Then he gave himself an electric shock by touching the bucket. The more violent the shock, the more electricity he had generated.

Today, we have various instruments for measuring electricity. We can safely use electricity in our everyday lives without worrying about getting a shock.

In this module we will find out about static electricity and how it is used. We will see how the idea of electric charge can explain what is going on in electric circuits. We will also look at how electricity is generated and sent around the country, to be used at the flick of a switch.

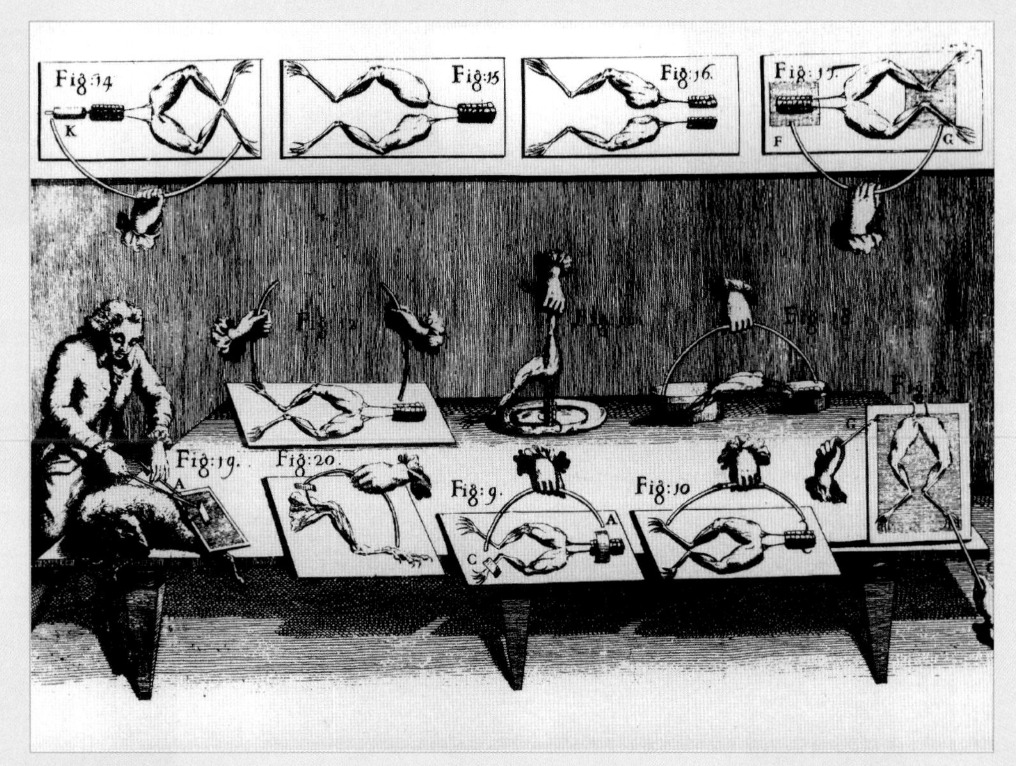

In the eighteenth century, scientists had unusual ways of showing when they had generated electricity. One way was to connect up to frog's legs. If the legs twitched, it showed that a current was flowing.

Before you start this module, check that you can recall the answers to the following questions about electricity.

1. What meters are used to measure electric current and voltage?
2. In what units are current and voltage measured?
3. How can you increase the current flowing through a component in a circuit?
4. What is the energy source in an electric circuit?
5. How can you make an electromagnet, and how can you increase the strength of its magnetic field?

10:1 Static electricity

The Van de Graaff generator in the photograph can be charged up to a very high **voltage** – hundreds of thousands of **volts**. If the voltage is high enough, sparks jump through the air.

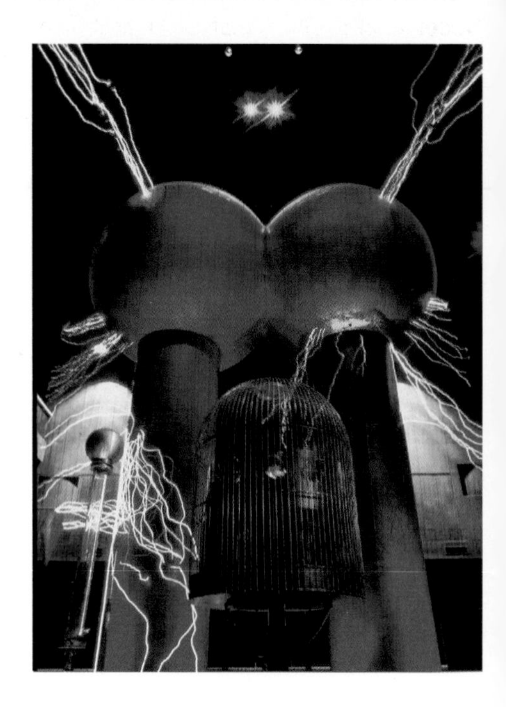

Charging up

If you rub a rubber balloon on your woollen jumper, they may both become charged. You can use the charged balloon to attract small pieces of paper, or your hair.

These materials – rubber, wool, paper, hair – are all electrical insulators. Two insulators may become charged by the force of friction when they are rubbed together. A charged material can attract an uncharged insulator.

Two types of charge

There are two types of **electric charge**, called **positive** and **negative**.

After it has been rubbed, the jumper has a positive electric charge. The balloon has a negative charge. They will attract each other because *opposite charges attract*.

The two charged balloons repel each other. Each has a negative charge, and charges that are the same repel one another (*like charges repel*).

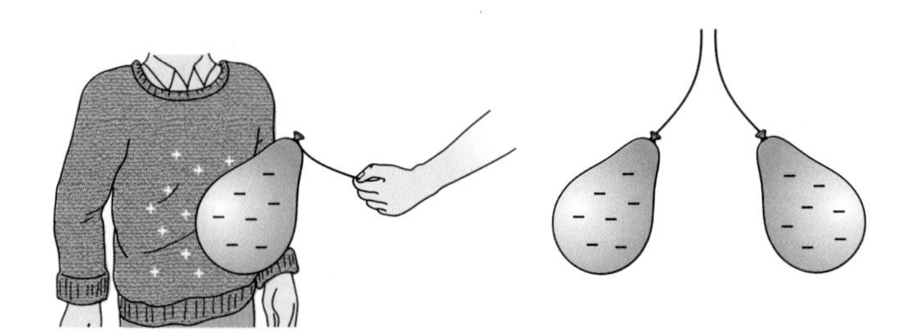

a Copy and complete the table, to show how these pairs of objects will affect one another.

First object	Second object	Attract/repel/no effect?
Negatively charged balloon	Negatively charged balloon	
Negatively charged balloon	Positively charged nylon thread	
Negatively charged balloon	Uncharged piece of wool	

Dangerous sparks

Static electricity can be dangerous. Sparks can give you a shock, or cause a fire.

When a tanker delivers petrol to a filling station, the petrol may become charged by friction as it flows along the pipe. A spark could then jump through the air, and this might ignite the petrol vapour.

To avoid this danger, the pipe is connected to the earth using a metal wire. Then the charge can escape safely into the ground.

This fuel store exploded when an electrical spark ignited the petrol vapour.

Van de Graaff generator

A school laboratory often has a Van de Graaff generator, a small version of the one shown in the photo on page 170. If you put your hand near the dome when it is charged, a spark may jump and give you a shock.

Connect the dome to the earth with a wire or rod. Then it will discharge safely.

b A wire or rod is used to discharge the charged dome. Explain why it should be made of metal.

Questions

1 What force causes two objects to become charged when they are rubbed together?

2

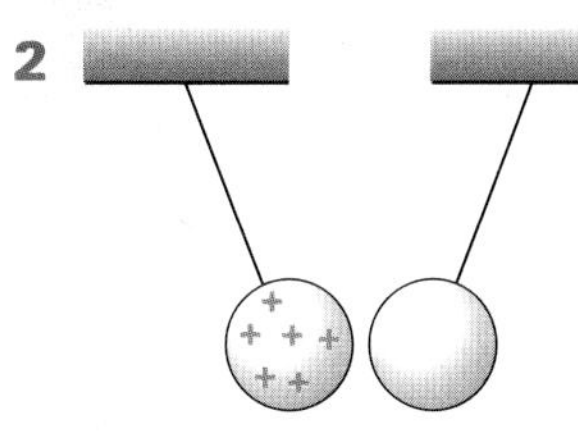

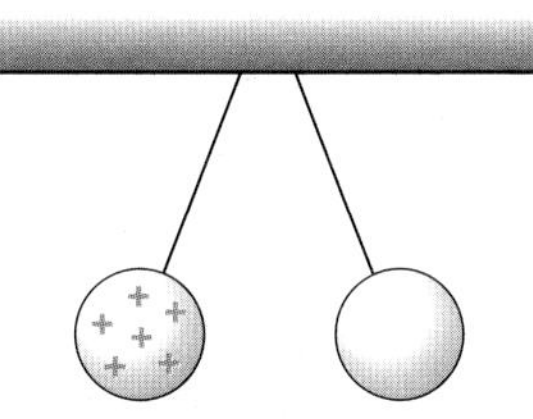

The diagrams above show charged balls hanging on strings. Copy the diagrams and complete them to show the charge (+ or –) on each ball.

3 Here is something you may have noticed:

- If you comb your hair, your hair is attracted to the comb.
- After combing, your hair is light and fluffy – your hairs are all repelling each other.

What do these observations tell you about the electric charges on hair and comb?

4 Copy the sentence below. Use words from the list to fill the gaps.

conduct **connect** **discharge**

To __________ a charged metal object safely, __________ it to earth using a metal wire so that the charge will __________ away.

Summary

- When two different insulating materials are rubbed together, they may become electrically charged.
- Two objects with the same type of charge will repel each other.
- Objects with opposite charges attract.
- To discharge a charged object, connect it to earth using a conductor.

10:2 Using static electricity

Static electricity can be useful. Some machines work using static electricity. The picture shows how static electricity helps when part of a car is spray-painted.

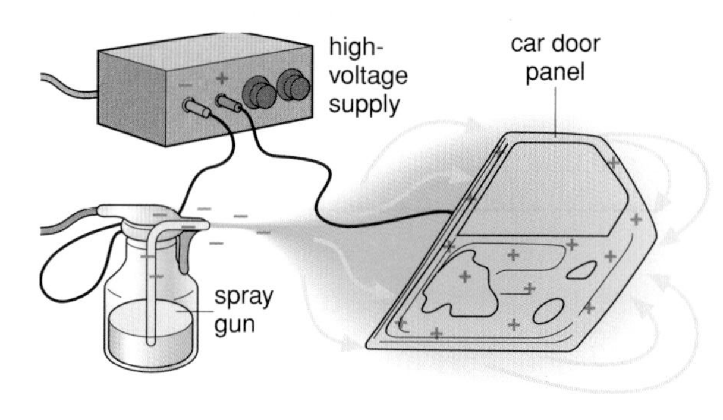

The droplets of paint have a negative charge. They are attracted to both sides of the positive body panel, so the panel is painted on both sides.

When static electricity is made use of, we are using the fact that two things which are charged can:

- attract one another (opposite charges), or
- repel one another (like charges).

a Explain why the paint is attracted to the metal panel.

Inkjet printing

The cans in the photo are printed with their 'best before' date and the batch number. The customer knows when it is safe to open the cans, and if there is a problem the whole batch can be recalled.

If you look closely, you will see that the letters and numbers are made of little dots of ink. The ink is sprayed onto the cans as they move along the production line. The inkjet printer makes use of static electricity to guide the droplets of ink. Fizzy drink bottles are also labelled like this. This machine can print on hundreds of cans or bottles every minute.

How the printer works

The diagram shows how an inkjet printer works. Look for:

the *gun* which produces the jet of ink droplets;

the *electrode* which charges them up;

the *deflecting plates* which move the jet from side to side.

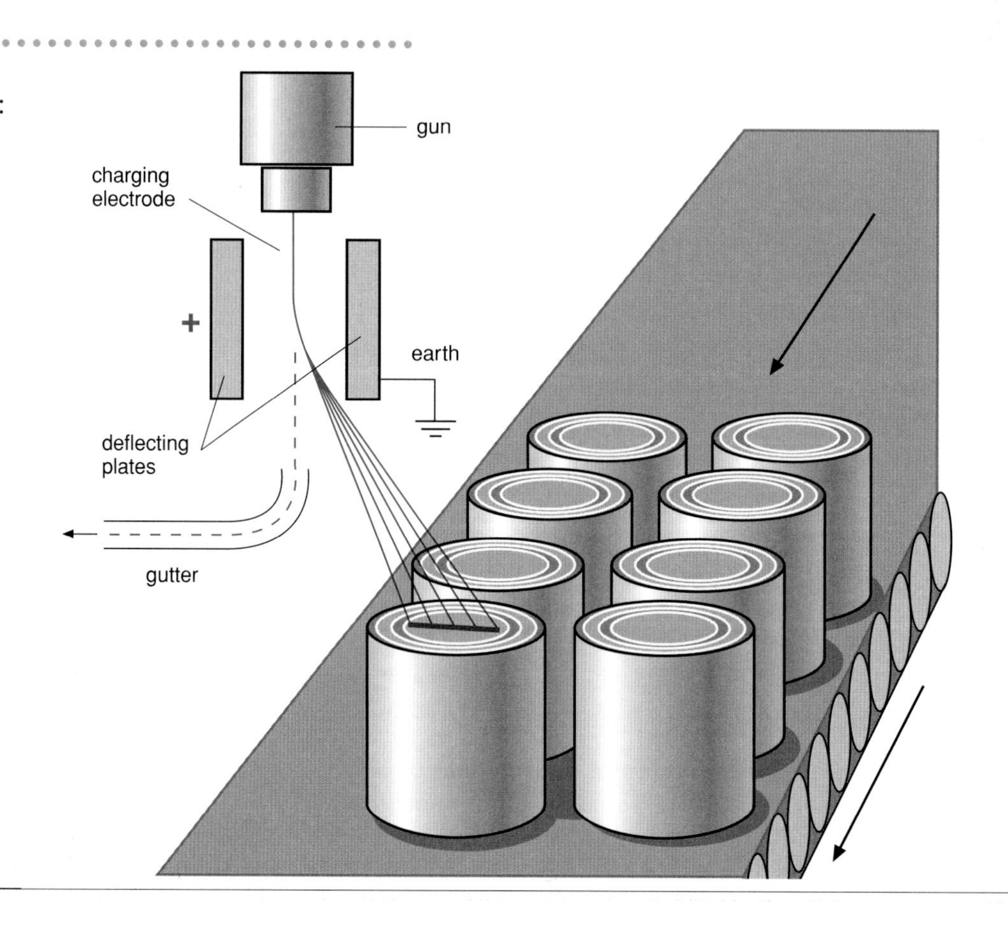

b Copy the following sentences; put them in the correct order to give an explanation of how an inkjet printer marks the cans.

- ◆ At the same time, the can moves underneath the jet so that the letters are traced out on it.
- ◆ The jet of charged droplets is moved sideways by the deflecting plates.
- ◆ The jet of tiny ink droplets is sprayed from a very fine nozzle. As it is sprayed, it is given an electric charge.
- ◆ The voltage between the plates is increased to deflect the jet further to the side.

c Do the ink droplets have positive or negative charge? How can you tell?

d How does the jet of ink droplets move if the deflecting plates have no charge?

> **Did you know?**
> Some inkjet printers scan the same area twice, to fill in the gaps between the tiny dots which make up the letters.

Computer printers

Some computer printers work in the same way as the inkjet printer. A very fine jet of ink droplets is guided by static electricity to form the letters and images on the page. There is a miniature version of the industrial can printer inside the printer's ink cartridge.

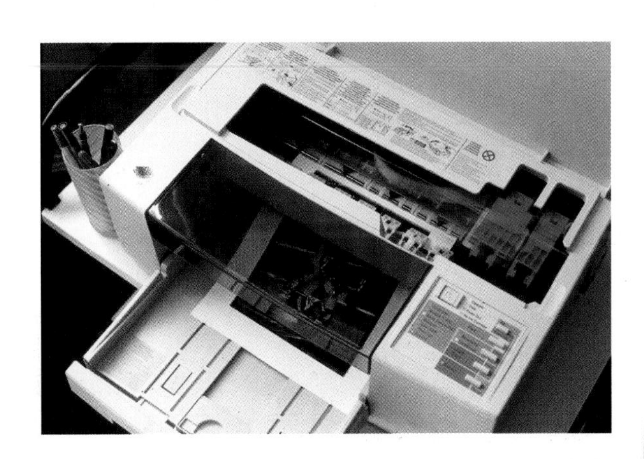

Questions

1 Copy and complete the following sentence, writing either **positive** or **negative** in the blank space.

If a drop of ink is given a positive charge, it will be attracted by a plate with a __________ charge.

2 Look at the diagram of the inkjet printing machine. Explain how the jet will move if the left-hand plate is given more positive charge.

Summary

- In inkjet printing, droplets of ink are electrically charged.
- The charged drops are directed to form the shapes of letters by the attraction and repulsion of charged plates.

10:3 Photocopier

A photocopier can make a copy of a black and white page in less than a second.

A bright light scans across the page to be copied. This makes an image on a special charged drum. The dark areas on the drum remain charged, and these attract the toner (ink). The toner is transferred to the paper.

a **Look at the diagram below. What charge does the drum have, positive or negative? What charge does the toner have? What charge does the paper have?**

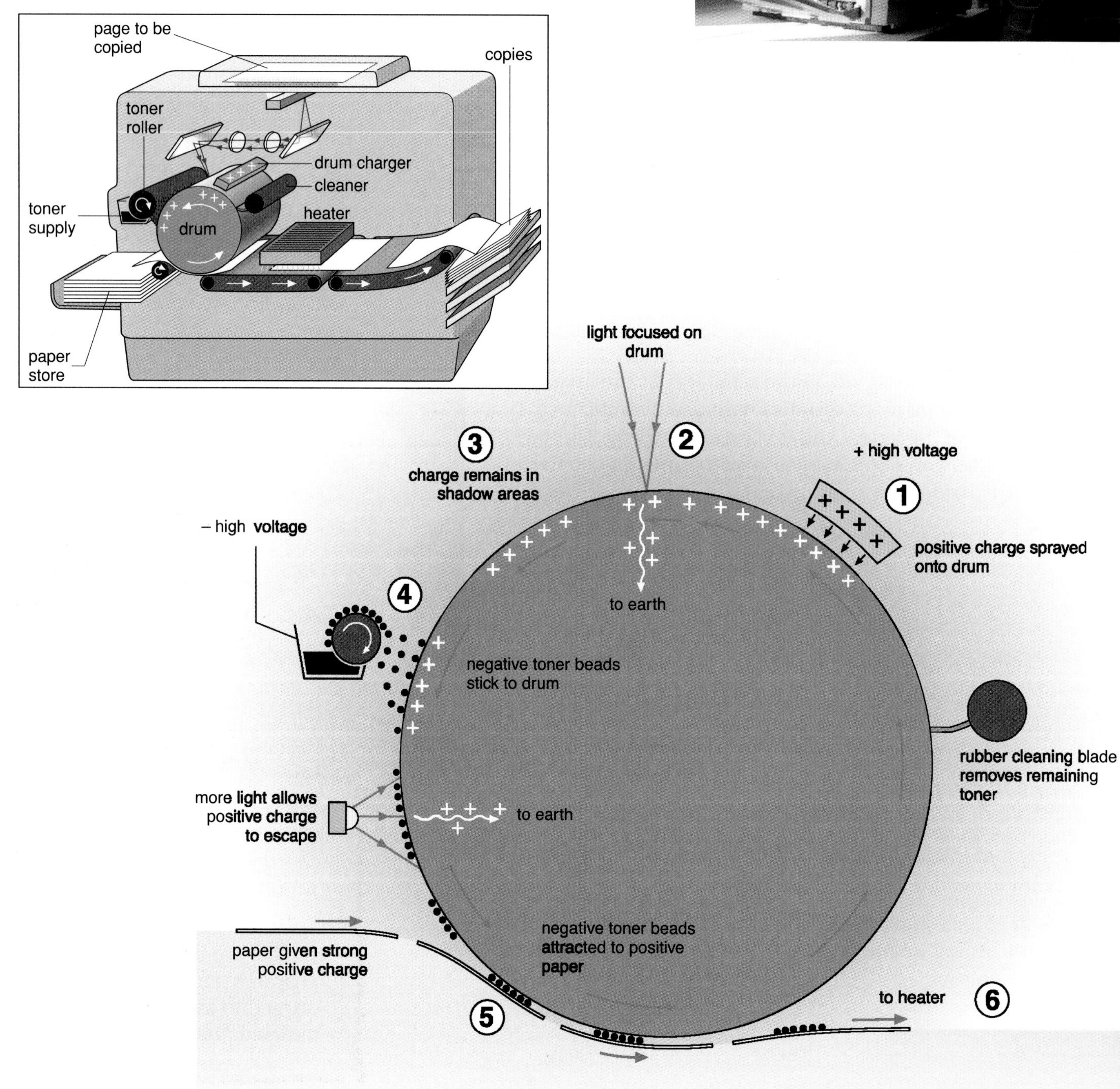

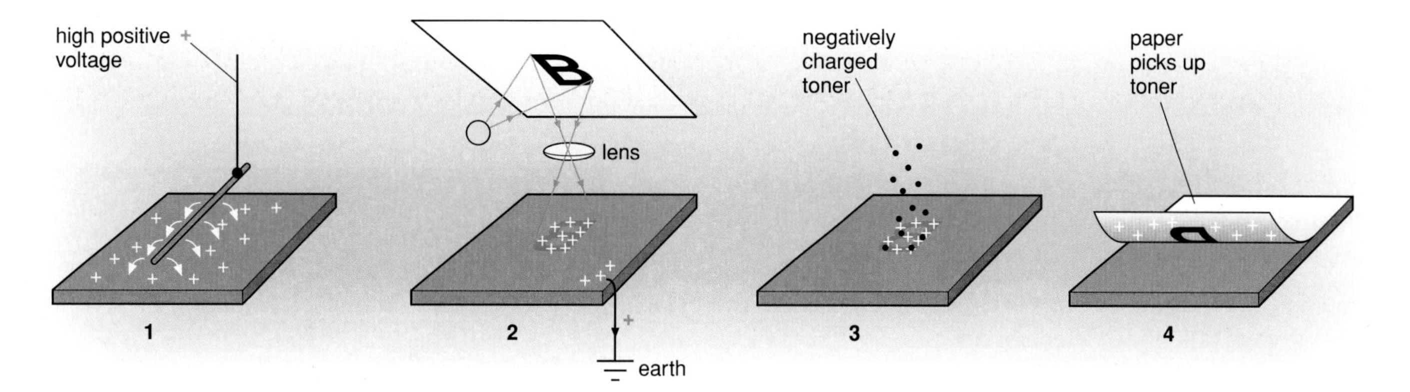

Did you know?
A laser printer works in a similar way to a photocopier. The laser beam provides the light which allows charge to flow away.

How it works

It is easier to understand the stages in making a copy if we picture the drum as a flat metal plate with a special coating. In the diagrams above:

1 Positive charge is sprayed onto the plate.

2 An image of the page is focused on the base plate:

- Where light falls, its energy allows the charge to escape to earth.
- Black print on the page leaves a shadow on the base plate, and the charge cannot escape.

3 Negatively charged black toner powder is attracted to the positive areas ofthe plate.

4 A sheet of positively charged paper picks up the toner powder.

The paper is then heated so that the toner powder melts and sticks to the paper.

b After the image has been focused on the base plate, what charge do the black areas have? What charge do the white areas have?

c Why is the toner powder attracted to the paper?

Stage number	What's happening
	The toner powder is attracted to the paper
	An image of the page is reflected onto the drum, allowing positive charge to escape
	Shadow areas on the drum keep their positive charge
	The toner powder is melted onto the paper
	The drum is given a positive charge
	Toner powder is attracted to the shadow areas on the drum

Questions

1 Copy and complete the following sentence, writing either **positive** or **negative** in the blank space.

Toner powder with a __________ charge is attracted by a positively charged sheet of paper.

2 Look at the diagram of the photocopier drum on the opposite page. Decide which stage in the process is described by each sentence in the table on the right. Then copy the table, putting the stages in the correct order.

Summary

- In a photocopier, light provides the energy for positive charge to escape from the charged drum.
- Positive electric charge attracts the black toner powder to the drum, and then to the paper.

10:4 Electrons on the move

We can understand static electricity by thinking about **electrons**. An electron is a tiny particle, much smaller than an atom. It has a negative electric charge.

In an electric circuit, electrons are moving around all the time. That's why we talk about 'electronics'.

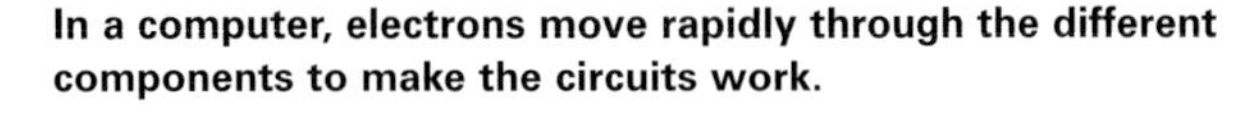

In a computer, electrons move rapidly through the different components to make the circuits work.

Getting charged

What happens when you rub a rubber balloon on your woollen jumper? Electrons are rubbed off the jumper onto the balloon.

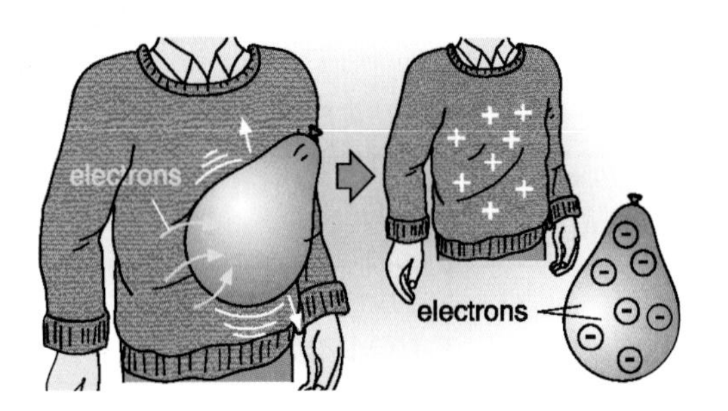

The balloon gains electrons, so now it has a negative charge. The jumper loses electrons, so now it has a positive charge.

It is easy to rub electrons off some materials. An electron is part of an atom. It is on the outside of the atom, so it is easily rubbed off.

a If you rub a polythene rod with a cloth, the rod becomes negatively charged. Which has gained electrons, the rod or the cloth? Which has lost electrons?

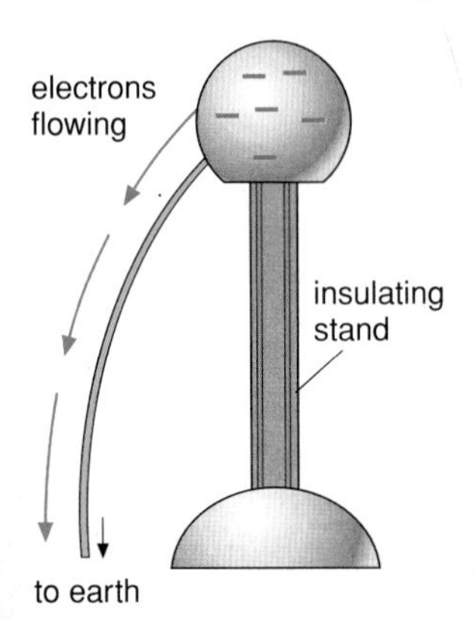

A negatively charged sphere has a lot of extra electrons. They flow down to the earth, through the wire.

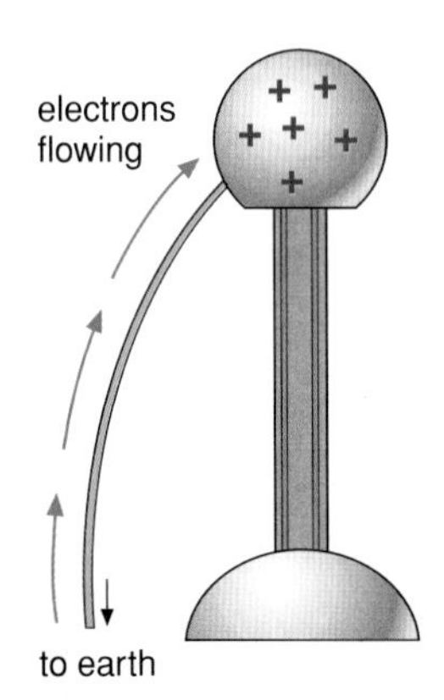

When a positively charged sphere is discharged, electrons flow up through the wire from the earth. The negative electrons cancel out the positive charge.

Electrons and current

This metal sphere has been charged up. To avoid sparks and shocks, it must be discharged.

When a wire is connected in a complete circuit, a current flows. If you could see inside the wire, you would see that the current is a flow of electrons.

> **Did you know?**
> It takes more than one million million million million million electrons to make a mass of one kilogram.

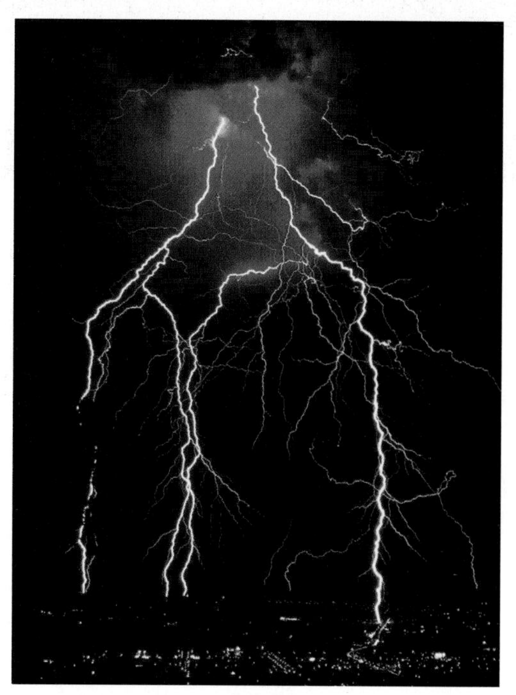

A lightning flash is a giant electric current. The thundercloud has a negative charge. When it discharges, electrons rush down to earth.

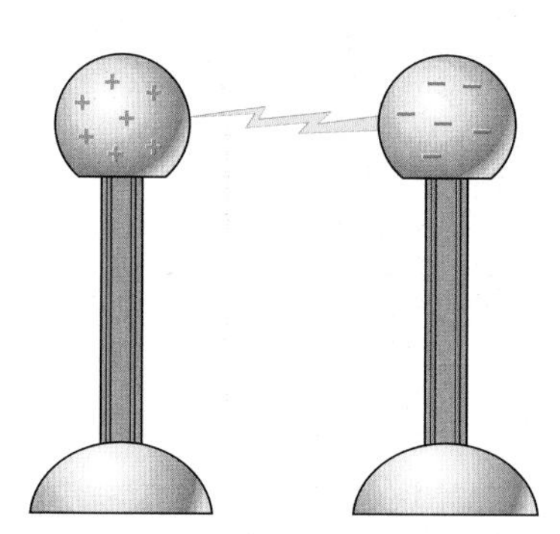

b The diagram shows two charged metal balls. A spark jumps between them. Copy the diagram, and add a labelled arrow to show which way the electrons jump through the air to make the spark.

Questions

1 Copy the sentences below, choosing the correct word from each pair.

Electrons are bigger/smaller than atoms.

Electrons have positive/negative charge.

Electrons are easy/difficult to rub from one material to another.

Electrons can move through metal/plastic.

2 A piece of fur is used to rub a plastic rod. Electrons are rubbed off the fur onto the rod.

Which of the following statements are true?

- The fur gains positively charged electrons.
- The rod gains a negative charge.
- The fur will repel the rod.
- The rod will attract the fur.

Summary

- Objects become charged because of the movement of electrons. Electrons have a negative charge.
- When electrons are rubbed from one material to another:
  - the material which gains electrons becomes negatively charged;
  - the material which loses electrons becomes positively charged.
- An electric current flowing in a solid conductor is a flow of electrons.

10:5 Electrolysis

Metals will conduct electricity, but most other solid materials won't. They are insulators.

There are two ways to make some of these insulators conduct. You need to turn them into liquids. Two rods called **electrodes** are used to connect the liquid to the battery.

- Method 1: melt the solid to make a liquid.
- Method 2: dissolve the solid in water to make a solution.

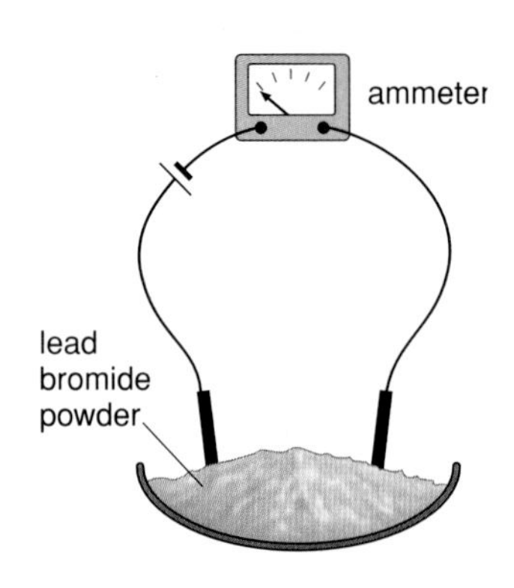

No current flows through the solid lead bromide powder.

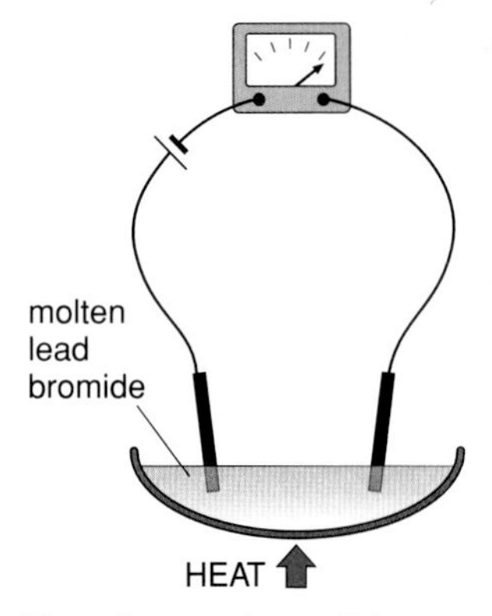

Heat the powder until it melts. Now a current will flow through it.

Melting a solid

a How can you tell from the diagrams that no current will flow through solid lead bromide?

b How can you tell that molten lead bromide is a conductor?

How the current flows

Compounds such as lead bromide are made of charged particles called **ions**. These ions can move about in a liquid. When copper chloride is dissolved in water, its ions become free to move about. The diagram shows how they move when a current flows through the solution.

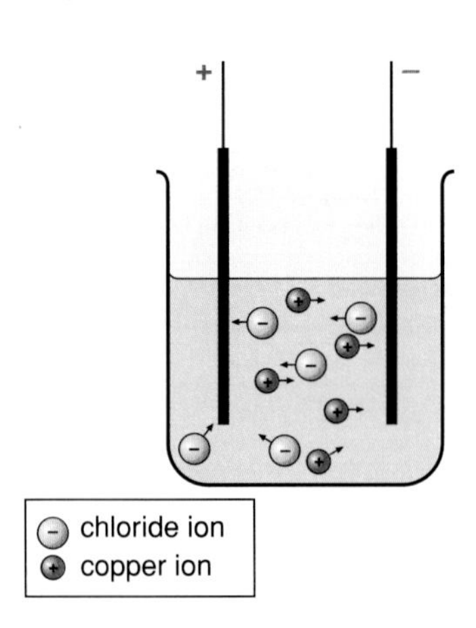

Copper ions have a positive charge, so they are attracted towards the negative electrode.

Chloride ions have a negative charge, so they are attracted towards the positive electrode.

c Look at the diagram which shows how the ions move when molten lead bromide conducts a current. Copy the following sentences, writing either **positive** or **negative** in the blank spaces.

Bromide ions have a __________ charge, so they are attracted towards the __________ electrode.

Lead ions have a __________ charge, so they are attracted towards the __________ electrode.

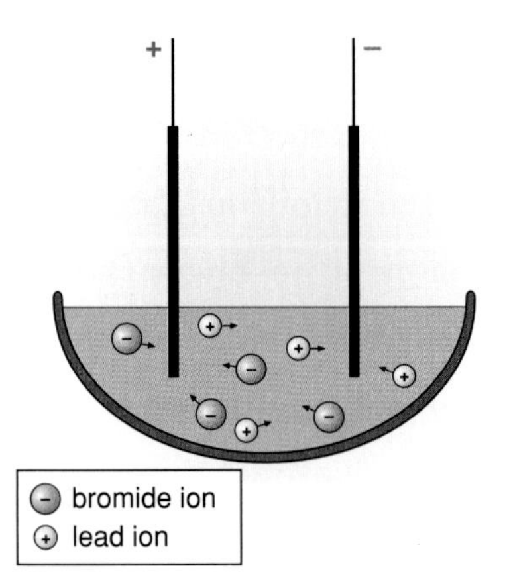

Splitting up

When a current flows through a liquid like this, the liquid is split up into simpler substances. This is called **electrolysis**. Electrolysis can be used to extract substances from their compounds. It can also be used to give a coating of metal to an object. Some jewellery is plated in this way.

compound	→	*simpler substances*
copper chloride	→	copper and chlorine
lead bromide	→	lead and bromine

When a current flows through molten lead bromide, solid lead and bromine gas are produced at the electrodes.

d When a current flows through copper chloride solution, what solid substance is coated on to the negative electrode? What gas is produced?

Questions

1 Copy the table. Fill in the first column with words from the list to match the definitions in the second column

electrolysis   ions   electrodes   electrons

Word	Definition
	Charged particles that move through a metal when a current flows
	Charged particles that move through a solution when a current flows
	Terminals used to make a current flow through a liquid
	Splitting up a liquid using an electric current

2 The diagram shows how an electric current can flow through lead iodide. Copy the sentences below, choosing the correct word from each pair.

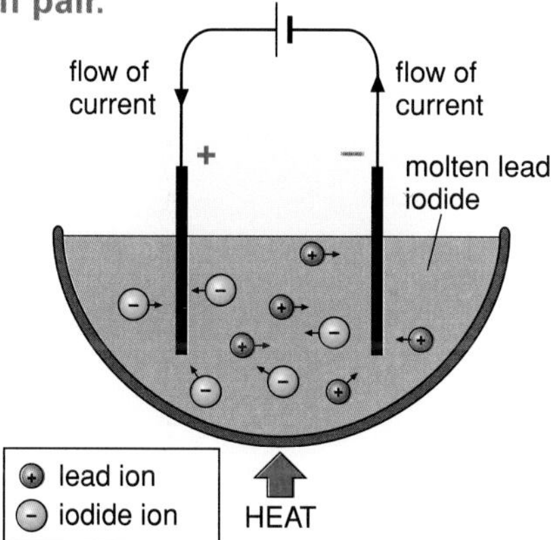

The lead iodide was melted/dissolved to make it conduct electricity.

Current enters the liquid through the positive/negative electrode.

Lead/iodide ions are attracted towards the negative electrode.

Lead/iodine is produced at the positive electrode.

Did you know?
If you get an electric shock, the liquid in the cells of your body is electrolysed.

Summary

- In electrolysis, a chemical compound is dissolved in water or melted.
- Negative ions move to the positive electrode; positive ions move to the negative electrode.
- Simpler substances are released at the electrodes.

10:6 Series circuits

We use electric circuits every day. Some, like a torch, have just a few components. The circuit diagram shows how they are connected together.

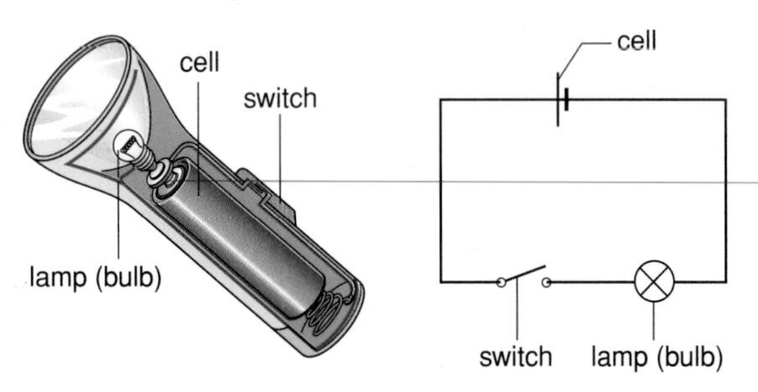

This torch has just three components, connected end-to-end (in series).

A computer is a complicated circuit, with many different types of component.

Some circuits, like those in a computer, have hundreds or thousands of components. When you know the rules for simple circuits, you can apply them to more complex circuits. The rules are the same.

a **Draw and label the circuit symbols for a cell, a filament lamp and a switch.**

Making a current flow

In a torch, the components are connected *in series* (end-to-end). The cell provides the voltage or **potential difference (p.d.)** needed to make a current flow. Some torches have more than one cell; two or more cells together make a **battery**.

A single cell provides a potential difference of 1.5 volts (1.5 V).

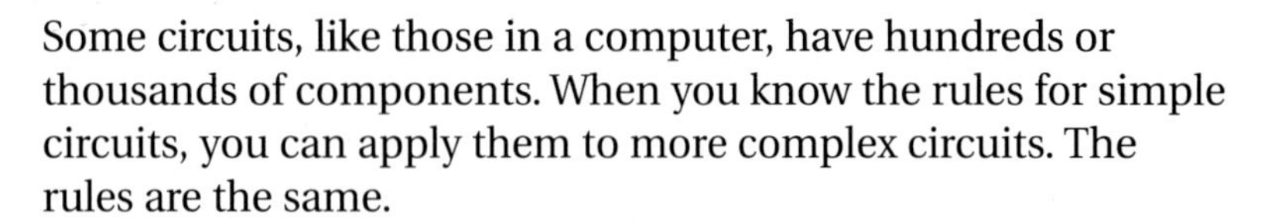

Cells are joined together to make a battery.

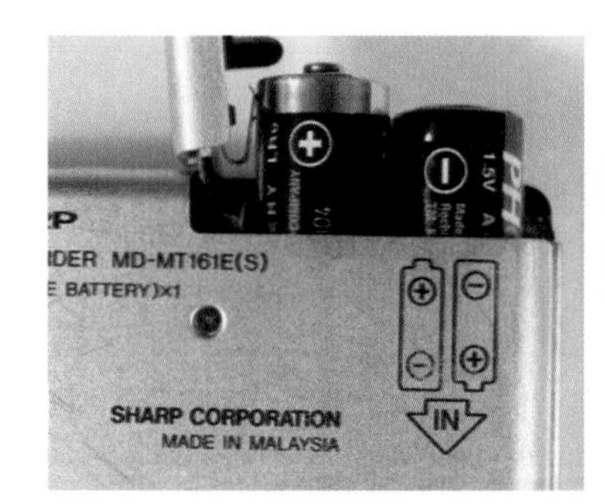

The diagram shows how to put the cells the right way round in this personal stereo.

When cells are connected in series, their p.d.s add up. Take care! If you connect a cell the wrong way round, the p.d. will be less.

b **What p.d. will be provided by each of these combinations of cells?**

(each cell 1.5V)

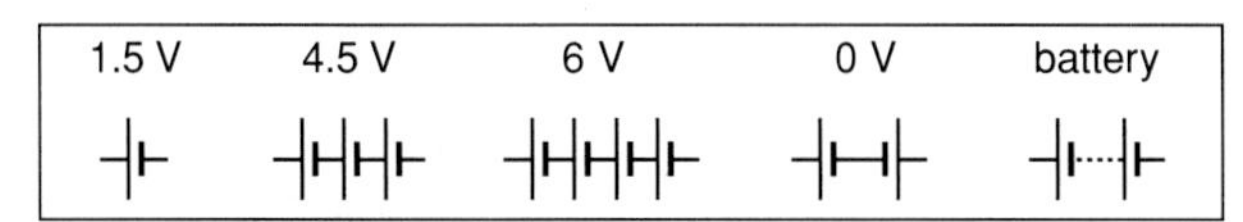

Did you know?

Rechargeable cells provide a smaller p.d. than disposable ones of the same size, because they contain different chemicals.

Resistors in series

This circuit has two resistors in it. It would be easier for the current to flow if there was only one resistor. There is more **resistance** in the circuit with two resistors. Resistance is measured in **ohms** (Ω).

Add up the resistances to find the total resistance in the circuit:

Total resistance = 10 Ω + 20 Ω = 30 Ω

A voltmeter can show the p.d. across a resistor. In this circuit:

- one resistor has a p.d. of 2 V across it;
- the other resistor has a p.d. of 4 V across it.

Together, these p.d.s add up to 6 V, the same as the p.d. of the battery. The p.d. of the battery has been shared between them.

c Two 50 Ω resistors are connected together in series. What is their combined resistance?

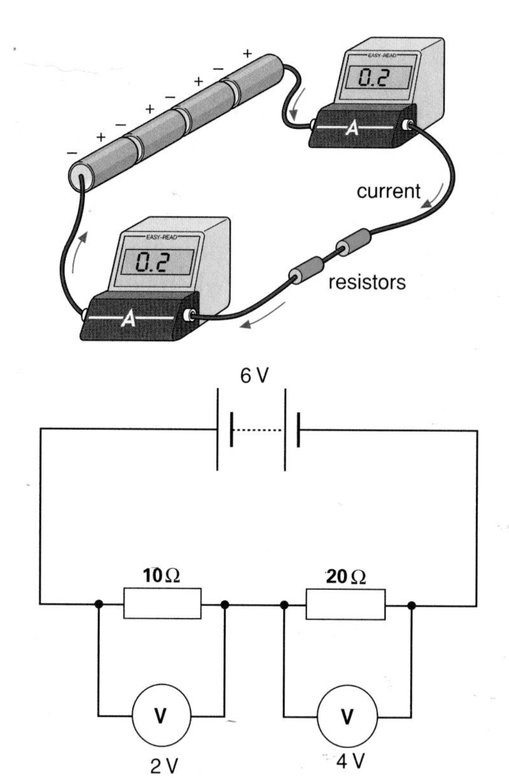

Rules for series circuits

- The same current flows through each component.
- The p.d. of the supply is shared between the components.
- Resistances in series add up.

Questions

1 Copy the circuit symbols on the right and label each with the correct name from the list below:

cell filament lamp resistor battery

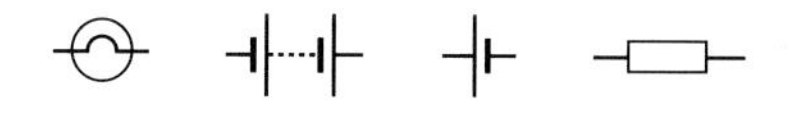

2 We use meters to make measurements in electric circuits. Copy the following sentences, choosing the correct word from each pair.

a An ammeter measures current/potential difference in amperes (A)/volts (V).

b A voltmeter measures current/potential difference in amperes (A)/volts (V).

3 Look at the circuit diagram below. Two resistors are connected to a battery of four cells. Copy the following sentences, choosing the correct word from each pair.

a The two resistors are connected in series/in parallel with each other.

b The four 1.5 V cells provide a p.d. of 1.5 V/6 V.

c The resistors have a total resistance of 10 Ω/30 Ω.

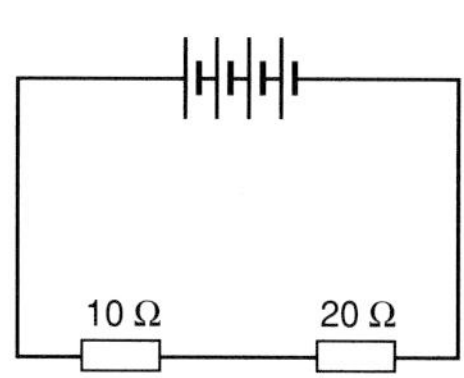

Summary

- When cells are connected in series, their potential differences add up.
- When components are connected in series (end-to-end):
  - the same current flows through each of them;
  - their resistances add up;
  - the p.d. of the supply is shared between them.

10:7 Components in parallel

Most lamps are labelled with the potential difference needed to make them light up brightly. It is important to choose the right lamp for the job. Then, when you connect it up to the right p.d., it will shine with the brightness you want.

What happens if you connect a lamp to the wrong p.d.?

- A smaller p.d. will make it dimmer, because a smaller current will flow.
- A bigger p.d. will burn it out, because the current flowing will be too big.

3 V torch bulb

12 V car headlamp bulb

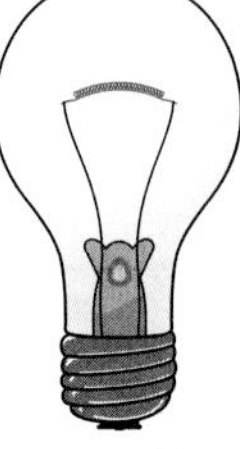

230 V mains lamp

Lighting two lamps

Here are two different ways to connect two lamps to a battery.

In series (end-to-end)

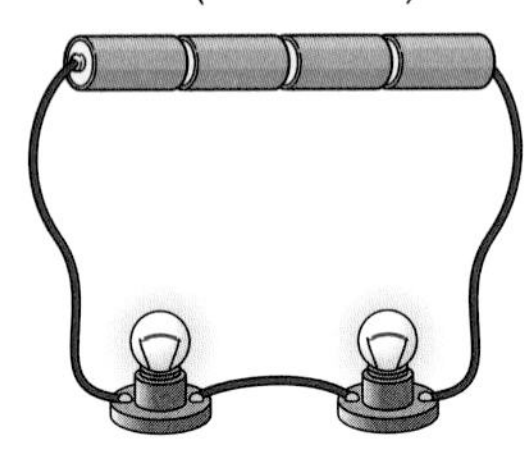

The p.d. of the battery is shared between the lamps. Each gets half the p.d., so they do not light up brightly.

In parallel (side-by-side)

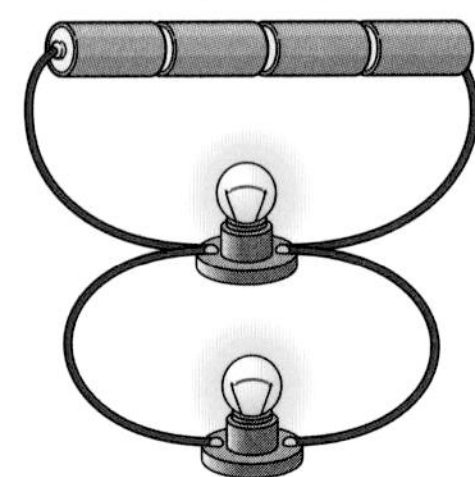

Each lamp gets the full p.d. of the battery, so each lights up brightly.

a **Which would be brighter, three lamps in series or three lamps in parallel?**

b **Draw circuit diagrams to represent the two circuits shown.**

How the current flows

The circuit diagram shows two resistors connected in parallel. A battery makes current flow round the circuit.

- The current flows out of the positive terminal of the battery.
- The current splits up. Some flows through one resistor, and the rest flows through the other resistor.
- Then the two currents join up again and flow back to the battery.

Because the two resistors are connected in parallel, the current is shared between them.

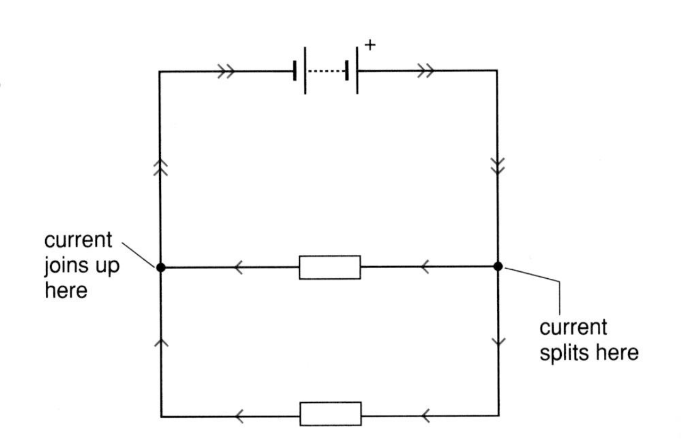

Measuring the current

Three ammeters have been included in this circuit.

- One measures the current flowing from the battery.
- The others measure the current flowing through each lamp.

c **Which ammeter measures the current flowing from the battery?**

d **What current will ammeter 3 show? Explain how you know.**

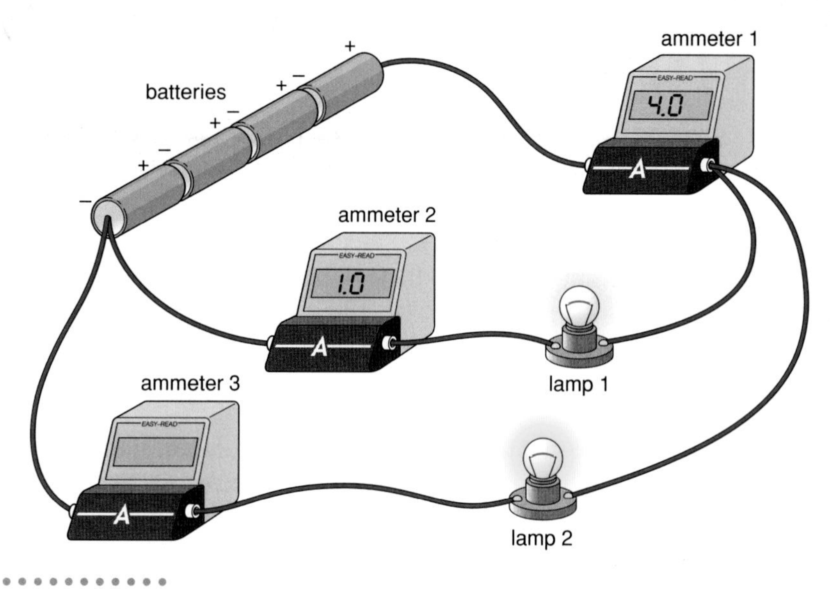

Sharing the current

Lamp 2 has a bigger share of the current flowing through it. It is easier for current to flow through lamp 2 than lamp 1. This tells us that it must have a lower resistance than lamp 1.

Rules for parallel circuits

- The p.d. is the same across each component.
- The current flowing from the supply is shared between the components.
- Components with less resistance get a bigger share of the current.

Questions

1 **Look at the picture of the circuit.**

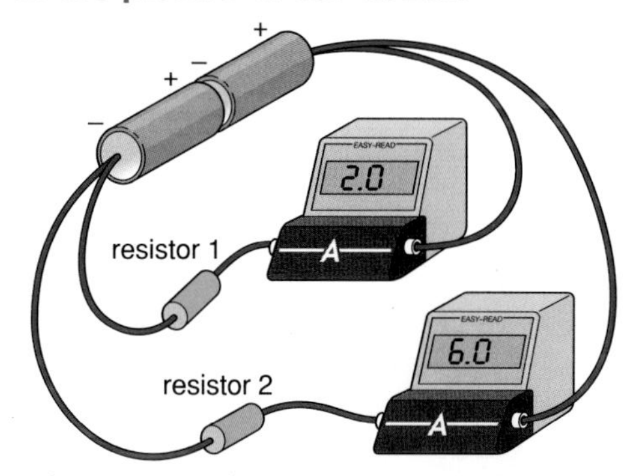

a **Are the two resistors connected together in series or in parallel?**

b **Which resistor has the greater resistance? How can you tell?**

c **How much current flows from the battery?**

2 a **Draw a circuit diagram to show two resistors, connected in parallel, in a circuit with a single cell.**

b **Add voltmeters to show how you would measure the p.d. across the cell and across each resistor.**

c **What can you say about the readings on the voltmeters?**

Summary

- When components are connected in parallel (side-by-side):
  - there is the same p.d. across each component;
  - the total current flowing around the circuit is shared between them;
  - a component with a greater resistance gets a smaller share of the current.

10:8 How many amps and volts?

Circuit designers use resistors in electric circuits. Each resistor in the picture is labelled with its resistance measured in ohms (Ω).

A resistor with a small resistance will let through more current than one with a greater resistance, for a given p.d.

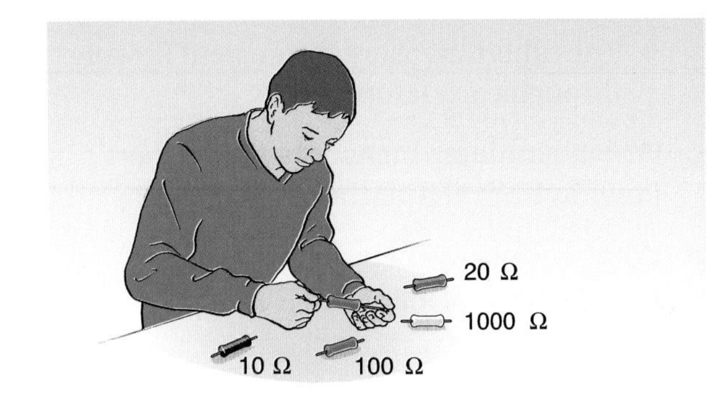

a Imagine that each resistor in the picture is connected up, in turn, to a battery. Which will let through the biggest current? Which will let through the smallest current?

Volts, amps and ohms

The circuit designer needs to know how much current will flow though a resistor in a circuit. The picture shows an experiment to find out how many amps of current flow through a resistor when it is connected to a supply.

- The voltmeter shows that the p.d. across the resistor is 20 V.
- The ammeter shows that the current through the resistor is 2 A.
- The resistor is labelled with its resistance, that is 10 Ω (ohms).

You can see that, to calculate the volts, you multiply the amps times the ohms. We can write this as a formula:

Potential difference = current × resistance

You might find it easier to remember this as:

Volts = amps × ohms

Example: The lamp in the picture has a resistance of 4 Ω. What p.d. is needed to make a current of 3 A flow through it?

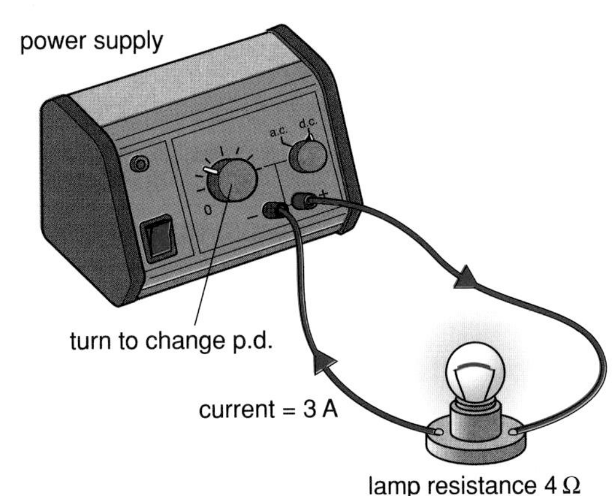

Step 1: What quantity are we trying to find? – The p.d.

Step 2: What quantities do we know?

current = 3 A resistance = 4 Ω

Step 3: Write down the formula and use it to calculate the p.d.:

Potential difference = current × resistance

Potential difference = 3 A × 4 Ω = 12 V

We know that the answer must be in volts, because amps × ohms = volts.

b What p.d. is needed to make a current of 0.5 A flow through a 20 Ω resistor?

Calculating current

A circuit designer needs to know how much current will flow through a resistor or other component. Here is the formula for working out the current:

Current = potential difference/resistance

Example: What current will flow through a 24 Ω resistor when there is a p.d. of 6 V across it?

Step 1: What quantity are we trying to find? The current.

Step 2: What quantities do we know?

Potential difference = 6 V

Resistance = 24 Ω

Step 3: Write down the formula and use it to calculate the current:

Current = potential difference/resistance

Current = 6 V/24 Ω = 0.25 A

So a current of 0.25 A will flow through the resistor.

c What current will flow through a 10 Ω lamp when there is a p.d. of 5 V across it?

Questions

1 Choose the two correct equations from the four below, and copy them out:

Current = potential difference × resistance

Current = potential difference/resistance

Potential difference = current × resistance

Potential difference = current/resistance

2 Copy and complete this equation, which shows how three units are related:

__________ = volts/ohms

3 A torch bulb has a resistance of 6 Ω. A current of 0.5 A flows through it when it is brightly lit. What p.d. is needed to make this current flow?

Summary

- To calculate the p.d. needed to make a certain current flow through a resistor, we use:

  Potential difference = current × resistance

  (volts = amps × ohms)

10:9 More volts, more amps

The sound technician moves the sliders to adjust the sound coming from each loudspeaker. Moving the slider up increases the voltage, more current flows through the loudspeaker, and the sound is louder.

You can change the brightness of a lamp by turning the knob on the power supply. When you increase the p.d., the lamp gets brighter. More current gets through.

a What happens to the brightness of the lamp when you decrease the p.d.? What happens to the current?

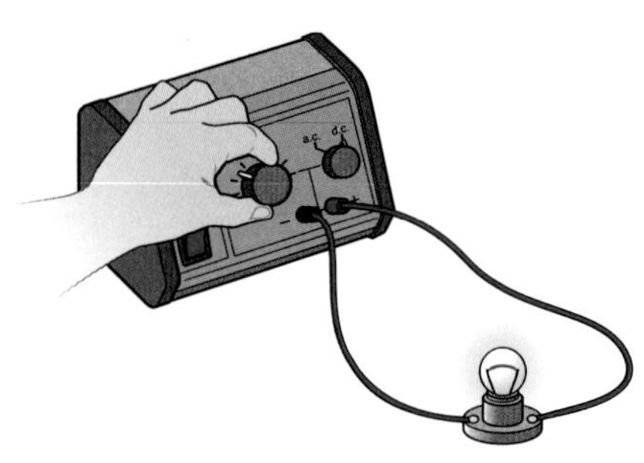

Increasing the voltage

How much current will flow through a resistor for a given voltage (p.d.)? You can use the circuit shown to find out.

- Turning the knob of the power supply changes the voltage.
- The meters show the voltage across the resistor, and the current flowing through it.

The graph shows that the current increases steadily as the voltage increases. The graph is a straight line.

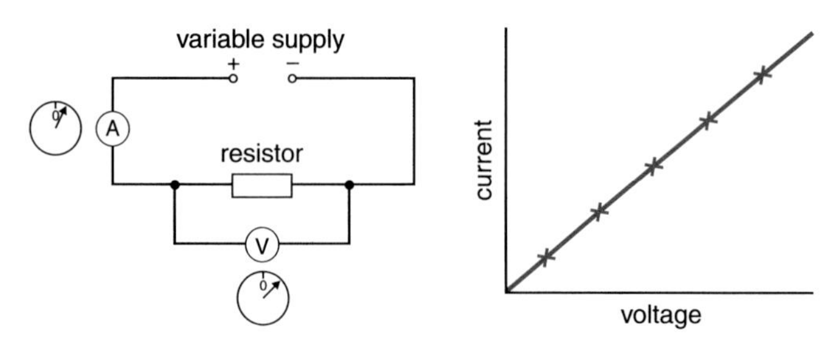

(This only works if the resistor doesn't get hot. A hot resistor would let through a smaller current.)

Current in reverse

Disconnect the power supply, turn it round, and clip it back again. This will make the current flow through the resistor in the opposite direction. The voltage and the current are now negative. Does this make a difference?

The graph shows that you get the same amount of current through the resistor when the voltage is reversed.

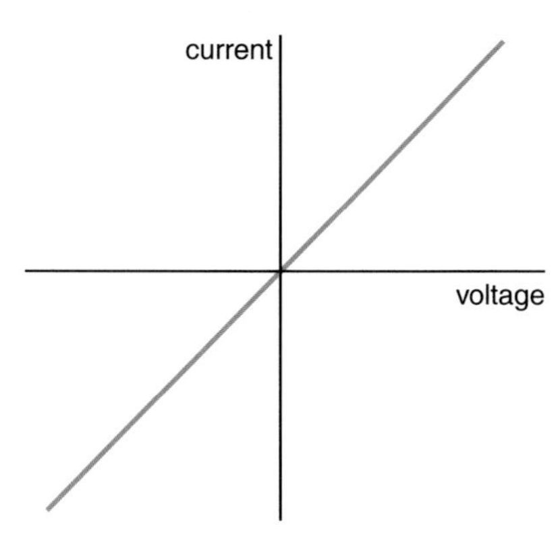

Resistor (at constant temperature).

b Copy this voltage–current graph for the resistor. Copy and complete this sentence to describe the graph:

When the p.d. across the resistor is increased in equal steps, the current through it also increases __________.

Light bulbs

A light bulb has a thin piece of wire called a filament. When the bulb is working, a current flows through the filament so that it glows. As the current through it increases, it gets hotter still. This makes it harder for the current to flow. The graph shows the effect.

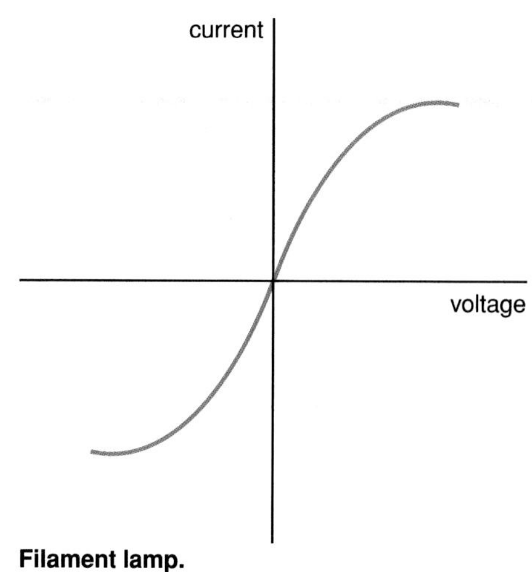

Filament lamp.

The current–voltage graph for a filament lamp is curved. As the p.d. across the lamp is increased in equal steps, the current through it increases in smaller and smaller steps. The resistance of the lamp is increasing.

c Draw a circuit diagram to show how you could make measurements of p.d. and current for a lamp.

Questions

1 The circuit in the diagram was used to see how the current through component X changed when the the p.d. was changed. The graph shows the results obtained. Copy the following sentences, choosing the correct word from each pair.

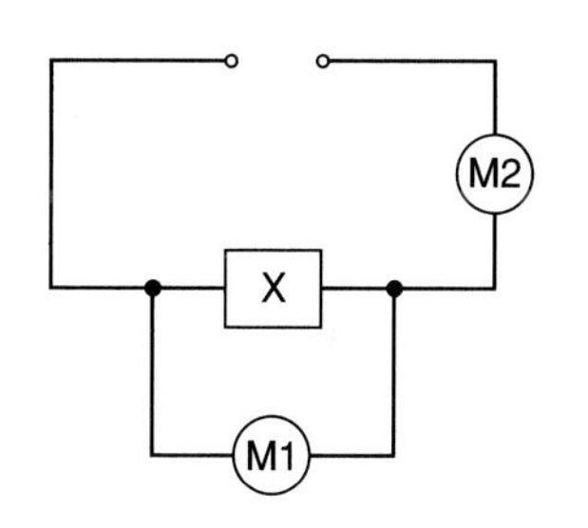

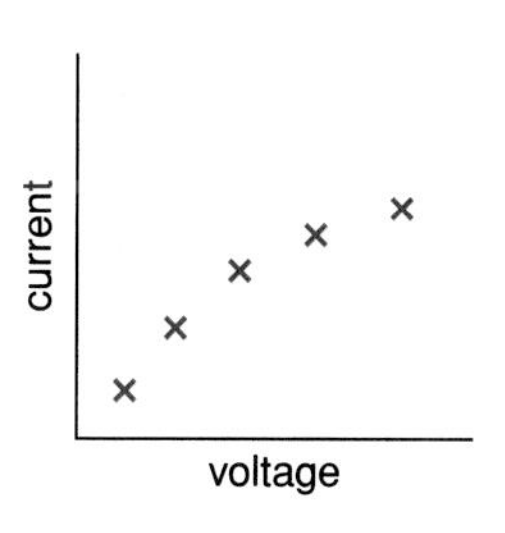

a Meter M1 is a voltmeter/ammeter.

b Meter M2 is a voltmeter/ammeter.

c The component X is a lamp/resistor.

d The graph shows that the current through X increases/decreases as the voltage across it increases.

2 The table shows how the current through a resistor changed when the voltage across it was changed.

Voltage (V)	Current (A)
0.0	0.0
2.0	1.6
4.0	3.2
6.0	4.8
8.0	6.4
10.0	8.0

a Draw a line graph to show this information.

b Use your graph to decide what current would flow through the resistor if the voltage was 5 V.

Summary

- The current through a resistor is proportional to the p.d. across it (provided the temperature remains constant, that is, stays the same).
- The resistance of a filament lamp increases as the temperature of the filament increases.

10:10 Three useful devices

A security engineer fits safety devices to people's houses, and to schools, offices and factories. How do these work?

- A fire alarm has a sensor which can detect a rise in temperature.
- A security light has a sensor which can detect when night falls.

Feeling the heat

A **thermistor** is a special type of resistor. It can be used as a temperature sensor.

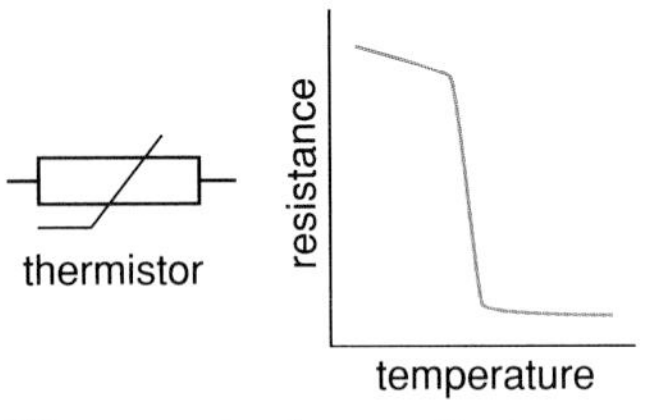

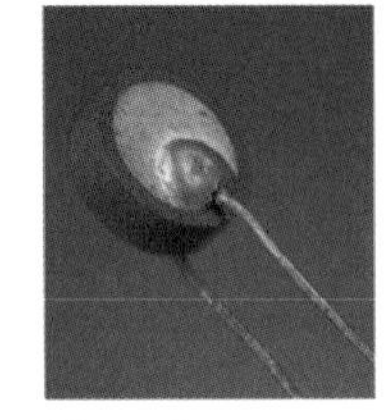
Thermistor

The graph shows that:

- when the thermistor is cold, its resistance is high;
- when the thermistor is hot, its resistance is much lower.

When the thermistor is hot, its electrons have more energy and it is easier for them to flow. The thermistor lets much more current flow. This current can make the fire alarm sound.

a Copy and complete this sentence, to explain how a thermistor works:

A thermistor lets through more current when it is hot because its resistance ___________.

Seeing the light

A **light-dependent resistor**, or LDR, is another special type of resistor. It can be used to detect light.

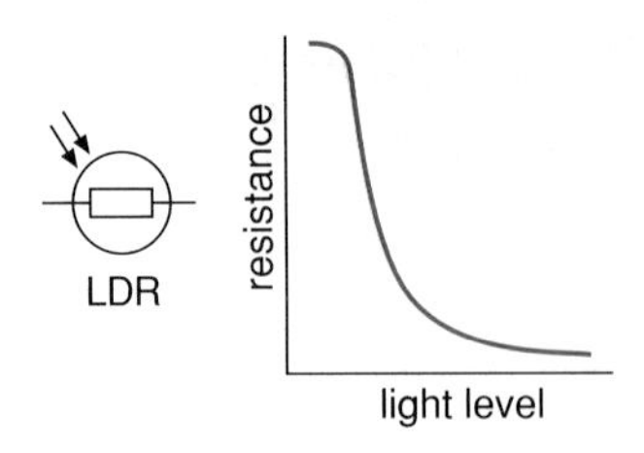

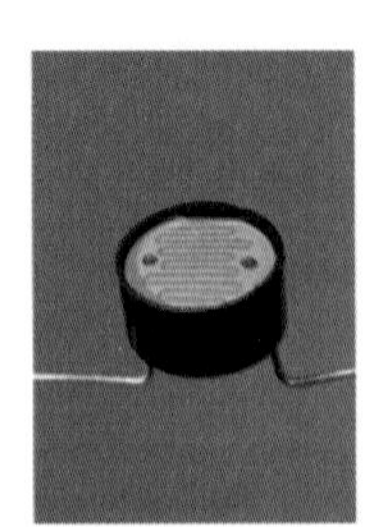
Light-dependent resistor

The graph shows that:

- when the LDR is in the dark, its resistance is high;
- when bright light shines on the LDR, its resistance is much lower.

The energy of the light helps the electrons to flow through the LDR.

This security light uses an LDR. At night, less light falls on the LDR. Its resistance increases. This switches on the light.

b Copy and complete this sentence, to explain how an LDR works:

When light falls on an LDR, it lets through __________ current because its resistance __________.

One-way traffic

A **diode** is another device used to control the current flowing in a circuit. It will only let current flow one way. If you connect it the wrong way round, no current will flow.

The circuit symbol shows which way the current can flow through the diode, in the direction of the arrow.

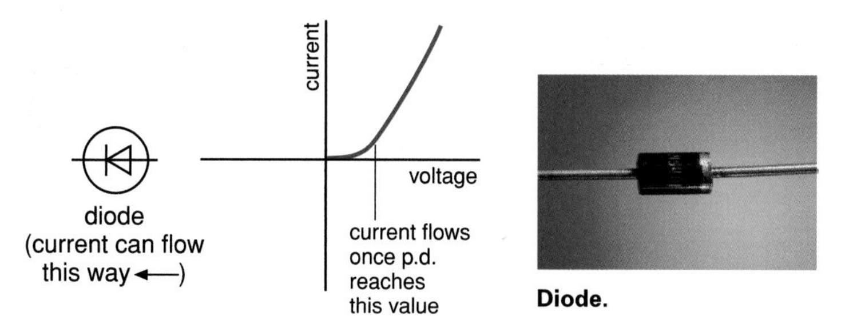

Diode.

The current–voltage graph shows that:

- current flows easily in the forward direction (once the p.d. is big enough);
- no current flows in the reverse direction; the diode's resistance is very high.

c How can you tell from the graph that a diode has a very high resistance in the reverse direction?

Questions

1 Copy these circuit symbols and label each with the correct name from the list:

diode **thermistor** **light-dependent resistor (LDR)**

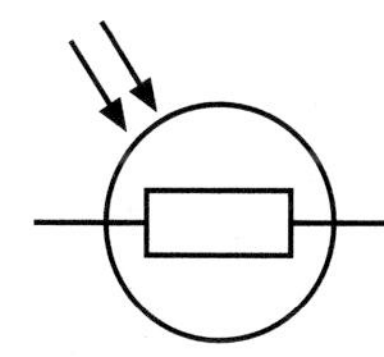
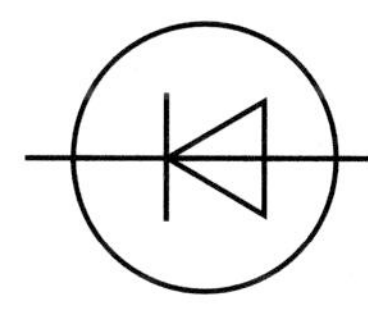
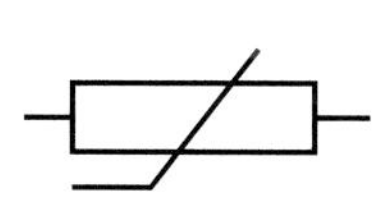

2 Copy these sentences, choosing the correct word from each pair:

a The resistance of a thermistor increases/decreases when it is heated.

b The resistance of an LDR increases/decreases when light falls on it.

c A diode will only allow current to flow in the forward/reverse direction.

Summary

- The resistance of a thermistor decreases as its temperature increases.
- The resistance of a light-dependent resistor (LDR) decreases as the light intensity increases.
- A diode allows current to flow in one direction only.

10:11 Wiring up safely

The mains electricity supply is dangerous. In the UK, its voltage is 230 V. If you touch the live wire, a current will flow through you, and it may be enough to kill you.

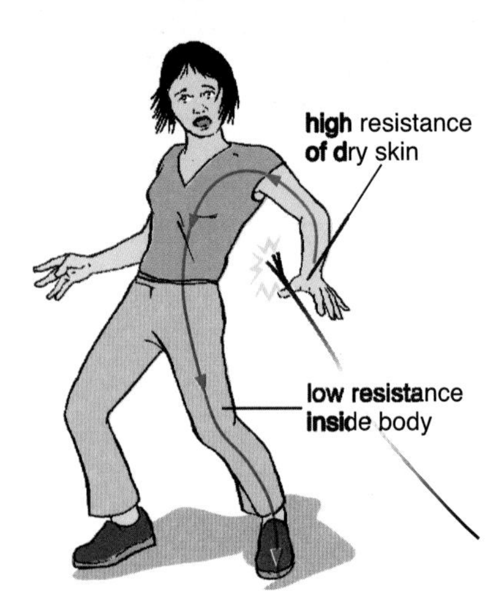

Dry skin has greater resistance than wet skin. This may be enough to save you from a fatal shock.

Plugs and cables

Today, most electrical appliances are sold with a plug already fitted to the end of the cable. This helps to ensure the user's safety.

Features of the cable:

- the metal wires are made of copper (a good conductor);
- the wires are surrounded by flexible plastic (a good insulator).

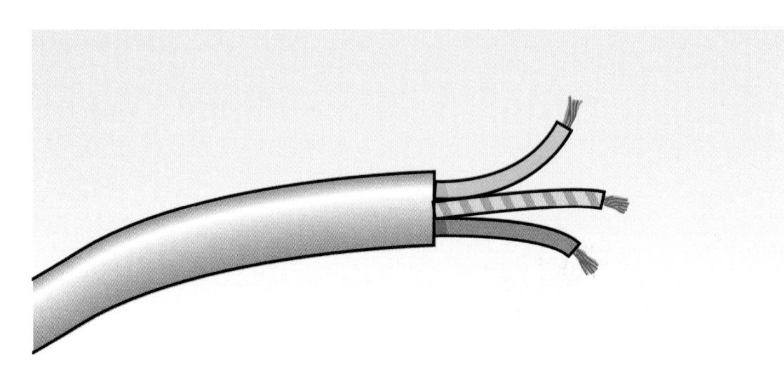

Did you know?
In the United States, the voltage of the mains is 110 V.

Features of the plug:

- the case is made of plastic or rubber (good insulators);
- the pins are made of brass (a good conductor);
- a fuse;
- an earth pin;
- a cable grip, to prevent the cable from being pulled out of the plug.

The diagram shows how a plug should be correctly wired.

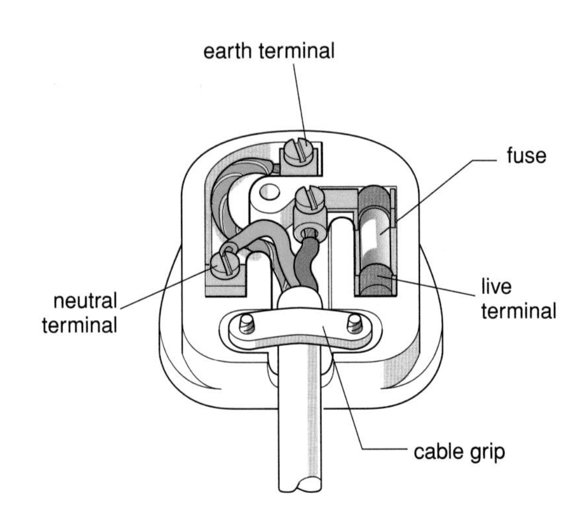

a Which two conducting materials are used in plugs and cables? Which two insulating materials?

b Copy and complete the table to show which wire goes to each terminal.

Colour of wire	Name of terminal	Position of terminal
brown		right
	neutral	left
		top

Things to look for

Plugs are sometimes known as '13 A plugs', because this is the biggest current they are designed to take. Look at a correctly wired 13A plug; you may notice some more safety features.

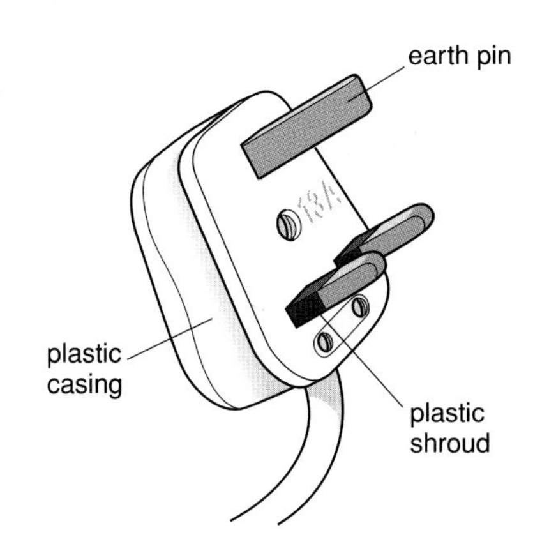

- The earth pin is the fattest and longest. It makes the first connection when you plug in, ensuring that the appliance is earthed before the live wire is connected.
- The live and neutral pins may have plastic shrouds, to protect your fingers if they curl round the plug as you push it into the socket.
- The bare wires are curved clockwise around the screws which hold them in place. As the screw is tightened, the wires are secured more firmly.

Questions

1 Which wire includes the fuse?

2 The drawings show plugs which have been incorrectly wired, or which are dangerous in some other way.

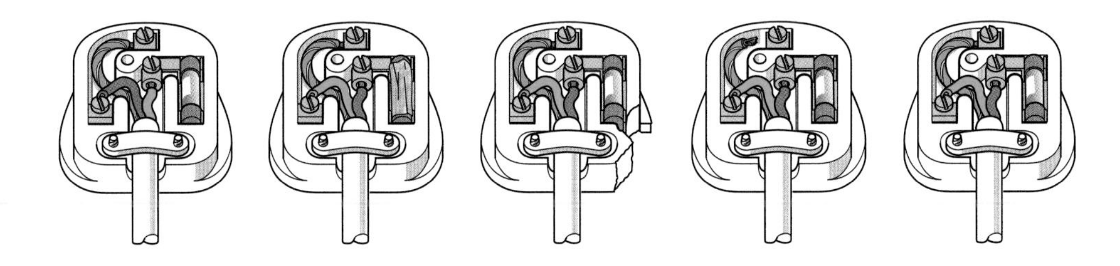

Match each drawing to the correct description in the list:

- **a** earth wire not connected
- **b** live and neutral wires swapped over
- **c** cable not secured by cable grip
- **d** case broken to expose live terminal
- **e** broken fuse replaced with aluminium foil

3 **a** Why is plastic a good material to use for the case of a 13 A plug?

b Why is brass a good material to use for the pins of a 13 A plug?

Summary

- The mains supply is 230 V; if not used safely, it can kill.
- Mains plugs and cables are designed to ensure the user's safety.

10:12 More safety features

The fuse is needed in the plug to ensure that the current flowing through the appliance does not become too big. Too much current can damage an appliance, or even cause a fire.

A fuse is labelled with its current rating and contains a small piece of wire. The current flows through the wire; if it exceeds the rating, the wire gets hot and eventually melts. This breaks the circuit. (A circuit breaker is sometimes fitted instead of a fuse – see page 196.)

The current rating of a fuse should be a little more than the normal current flowing through the appliance.

This electric fan caught fire as a result of an excessive current flowing through it.

These fuses are rated at 30 A, 13 A and 3 A.

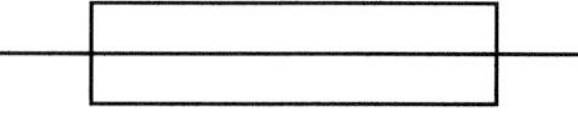

Circuit symbol for a fuse

a Choose fuses from those in the picture which would be suitable for:

a lamp, normal operating current 0.4 A;

a heater, normal operating current 8.7 A.

Earthing

Appliances such as washing machines and heaters often have metal cases. These can be hazardous if the live wire becomes connected to the case. The case should be connected to the earth wire.

b Explain why a 13 A fuse will not protect you if you get a shock from an unearthed appliance.

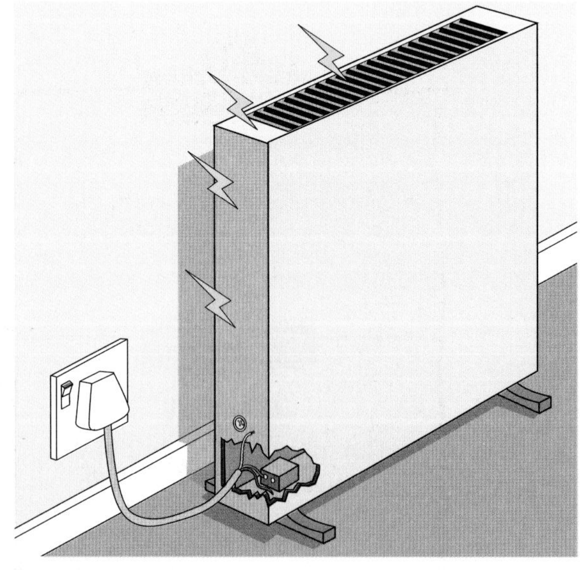

No earth connection:

Touch the live case and you will get a serious shock – possibly fatal. A current of just 0.1 A is enough to kill you.

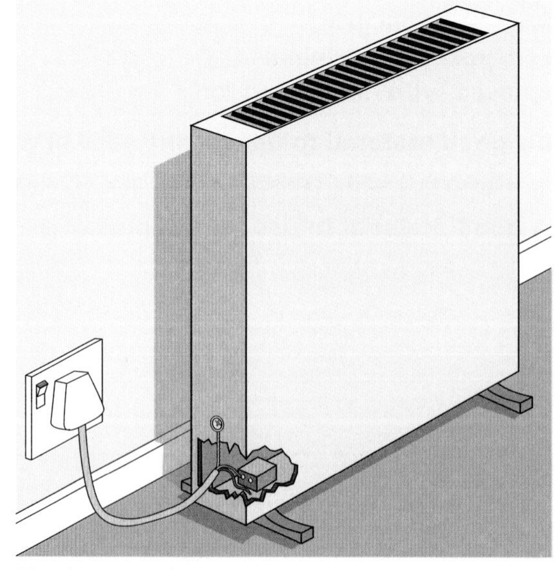

Metal case earthed:

A big current flows from live to earth. The fuse blows and you are safe.

A.c. and d.c.

Cells and batteries supply **direct current (d.c.)**, which flows steadily in one direction. Mains electricity is **alternating current (a.c.)**. Its direction keeps reversing. The frequency of the mains in the UK is 50 hertz (Hz), or 50 cycles per second.

c Which of the traces in the diagram shows a d.c. voltage?

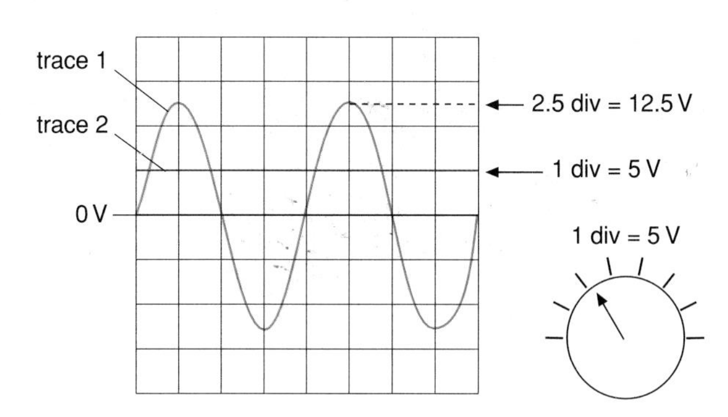

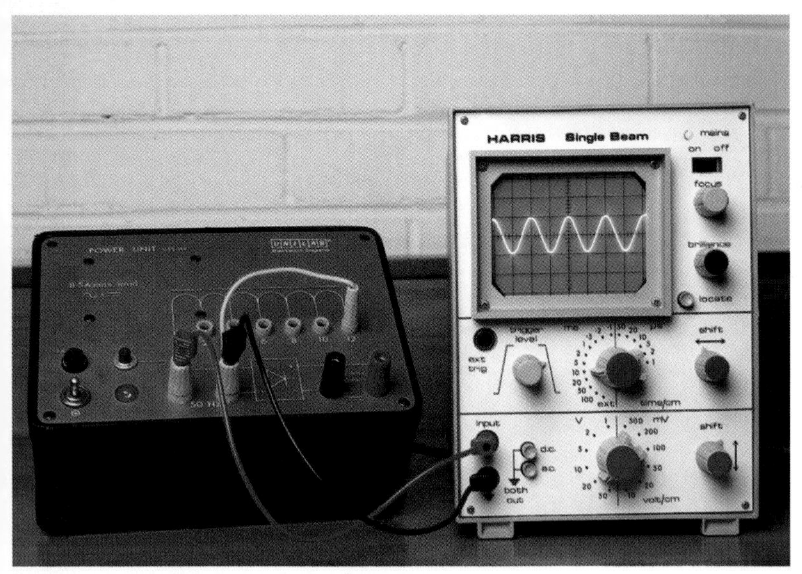

With a.c., the voltage goes up to a positive peak, then down to a negative trough.

An oscilloscope can show the difference between a.c. and d.c. From the trace, we can find the values of voltages.

Did you know?

A cable beneath the English Channel brings d.c. electricity from France. It has to be converted back to a.c. in the UK.

Questions

1 **Which two of the following can prevent an excessive current from flowing through an appliance?**

circuit breaker earth wire live wire fuse

2 **Copy the table and use terms from the list to fill the first column.**

230 V 50 Hz a.c. d.c.

	supplied by batteries
	supplied by the electricity mains
	the mains voltage
	the mains frequency

Summary

- The fuse in a plug ensures that the current flowing to an appliance does not become too great.
- Appliances with metal cases should be earthed.
- The mains supply is a.c.: it alternates with a frequency of 50 Hz (50 cycles per second).
- Cells and batteries supply direct current (d.c.).

10:13 Electrical power

We use electricity because it is a very convenient way of transferring energy from place to place.

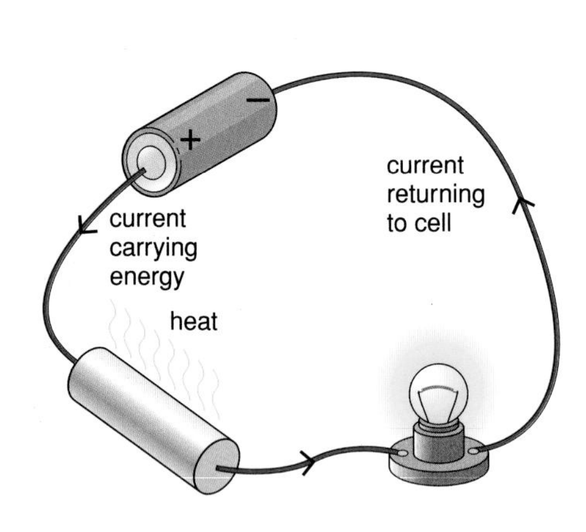

Energy is transferred from the cell by the charge as it flows around the circuit.

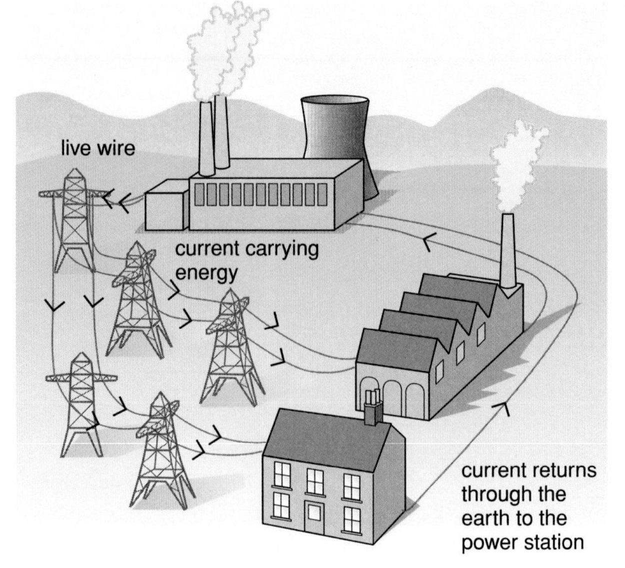

Energy is transferred from the power station to houses, factories etc.

Electrical energy can be easily changed into other forms:

- in a resistor: electrical energy → heat energy
- in a motor: electrical energy → kinetic energy
- in a lamp: electrical energy → light and heat energy

a When an electric current flows through a fuse, the fuse wire gets hot. What energy transfer is going on?

Power

The **power** tells us the rate at which energy is being transferred in a circuit. Power is measured in **watts (W)**. A power of 1 watt means that 1 joule is being transferred every second. (If you have studied the Energy module, you will know this already.)

1 watt = 1 joule per second 1 W = 1 J/s

b A light bulb has a rating of 100 W. How many joules of energy does it transfer each second, when working normally?

The power transferred by a circuit depends on the voltage and the current:

Power (W) = potential difference (V) × current (A)

You may prefer to remember this as a units equation:

Watts = volts × amps

Example: What power is transferred when a p.d. of 10 V makes a current of 2 A flow?

We have to multiply the volts times the amps:

Power = p.d. × current = 10 V × 2 A = 20 W

So the power is 20 W.

c The p.d. across a heater is 230 V, and a current of 2 A flows through it. What is the rate of energy transfer?

Did you know?
Current reaches your house through cables, but it flows back to the power station through the ground.

Calculating current

We can write the power equation in a different way, and use it to find the current flowing:

Current = power/p.d. amps = watts/volts

Example: A 100 W lamp works from the 230 V mains. What current flows through it?

We have to divide the watts by the volts:

Current = power/p.d. = 100 W/230 V = 0.43 A

So a current of 0.43 A flows.

This can be useful for calculating a suitable value for a fuse. Remember that the fuse rating must be a little higher than the normal operating current.

d What current flows through a 1000 W heater connected to the 230 V mains? Would a 3 A fuse be suitable to protect the heater?

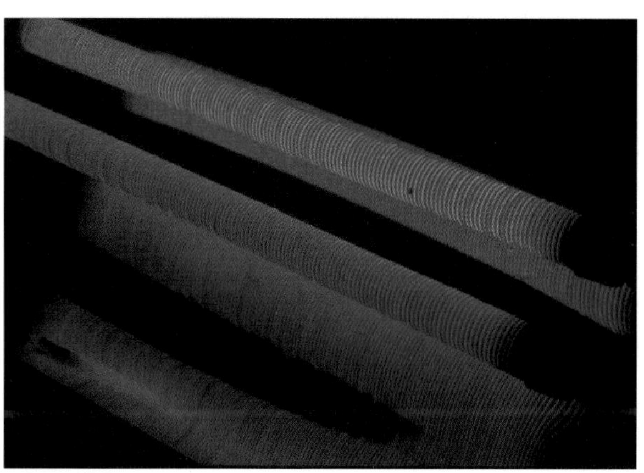

Each bar of this electric fire transfers energy at a rate of 1000 watts.

Questions

1 Copy the table. Choose words from the list to fill the first column.

energy charge power heat

	Flows round a circuit to make a current
	Transferred by an electric current in a circuit
	The energy transferred by a resistor
	The rate at which energy is transferred

2 Which two of these equations are correct?

p.d. = power × current

power = p.d. × current

current = power/p.d.

current = power × p.d.

3 A 1.5 V cell makes a current of 0.2 A flow through a torch bulb. What power is transferred?

Summary

- Electric current is a flow of charge, transferring energy around a circuit.
- As a current flows through a resistor, energy is transferred as heat.
- Power, the rate of transfer of energy, is given by:

  Power (watt, W) = p.d. (volt, V) × current (ampere, A)

10:14 Using electromagnets

Every school science lab has circuit breakers. They protect you when you are using electrical equipment. They use electromagnets to help make sure you don't get an electric shock. (An electromagnet is a coil of wire. When a current flows through the coil, it becomes magnetised.)

Trip switches

A trip switch is one type of circuit breaker. It switches the power off if the current gets too high. (It does the same job as a fuse.)

A power supply may have a 'trip' like this. If there is a fault in a circuit, the current may get too high and damage the power supply. When you have sorted out the circuit, you press the reset button to re-close the switch.

a Why does a small current not open the switch?

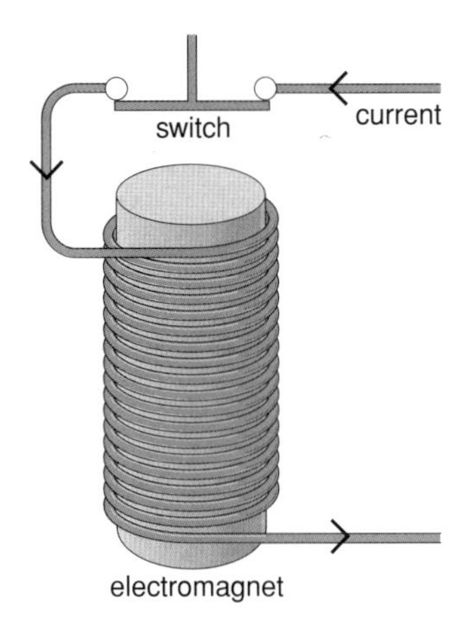

Inside the trip switch, the current flows through an electromagnet coil.

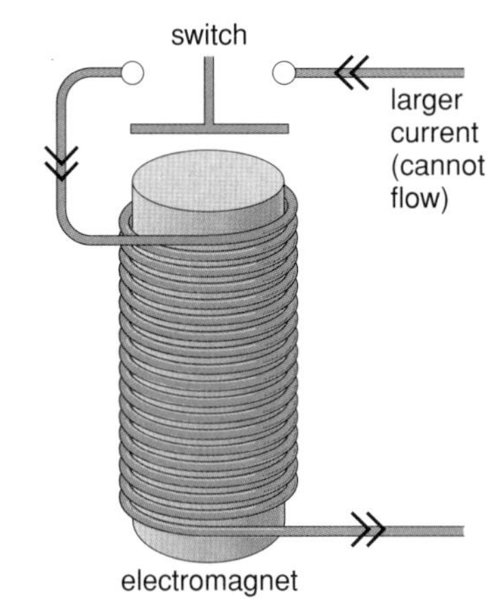

If the current is too big, the electromagnet pulls harder and the switch opens. The circuit is broken, so the current stops flowing.

Electric motors

Some cars run on electricity. Each wheel has an electric motor to turn it. Electric motors use electromagnets.

Petrol-driven cars also have electric motors in them. They have a starter motor to get them going, and motors for the windscreen wipers. There are motors to work the fuel pump and the screen washers, and there may be electrically operated windows and locks.

Parts of a motor

It is easiest to understand how a motor works by looking at a model. These are its important parts:

- a *coil of wire*, which becomes an electromagnet when current flows through it;
- two *permanent magnets*, to make a strong magnetic field;
- two *brushes* which touch onto the wire of the coil.

An electric current flows through one brush into the coil; around the coil; then back out through the other brush.

magnets
coil of wire
brush

How it works

When the current flows through the coil, it has a north pole and a south pole. Its north pole is attracted by the south pole of the permanent magnet, and so the coil starts to turn.

When the coil has turned halfway round, the brush connections swap over. Now the other side of the coil becomes the north pole, and is attracted round by the south pole of the permanent magnet. In this way, the coil keeps turning.

b The coil also has a south pole. Which pole of the permanent magnet is it attracted by?

Did you know?
Many electric motors have three coils inside them, instead of just one. They get pushed around in sequence, so that the motor turns more smoothly.

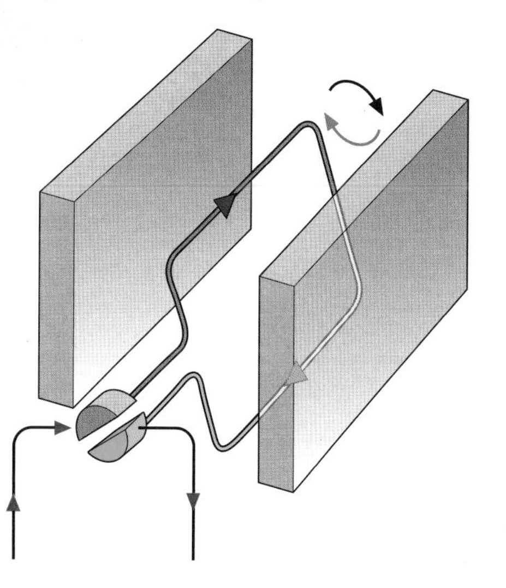

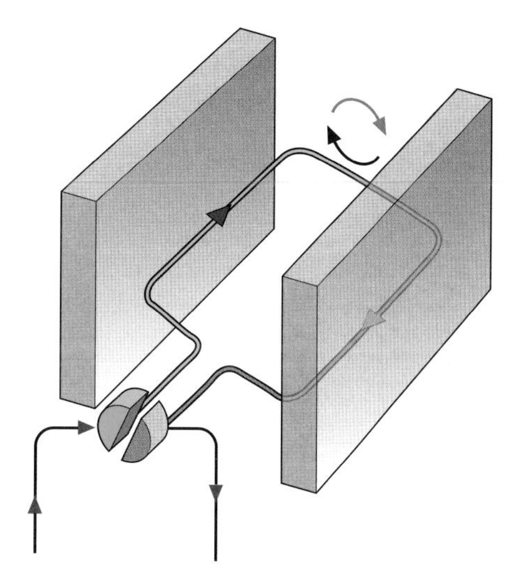

Questions

1 Read the sentences below and decide which word from the list should fill each gap. Then copy out the sentences in the correct order, to explain how an electric motor works.

magnets brushes coil

The coil is attracted by the __________ so that it rotates.

Current flows into the coil through the __________.

When the current flows round the __________, it becomes an electromagnet.

2 Explain why an electromagnetic trip switch is more convenient for the user than a fuse.

Summary

- Electromagnets are used in circuit breakers and electric motors.
- In an electric motor, a coil (an electromagnet) spins between two magnets, attracted round by the poles of the magnets.

10:15 More about motors

This strange-looking electric motor was invented by Michael Faraday. It was the world's first electric motor.

A copper wire hangs freely, with its end in a dish of mercury. There is a permanent magnet in the dish.

The battery makes a current flow down through the wire and then through the mercury.

There is a magnetic field around the current in the wire. It is repelled by the magnetic field of the permanent magnet. This pushes the wire round.

a If the current through the wire is increased, the motor turns faster. What does this tell you about the force pushing it?

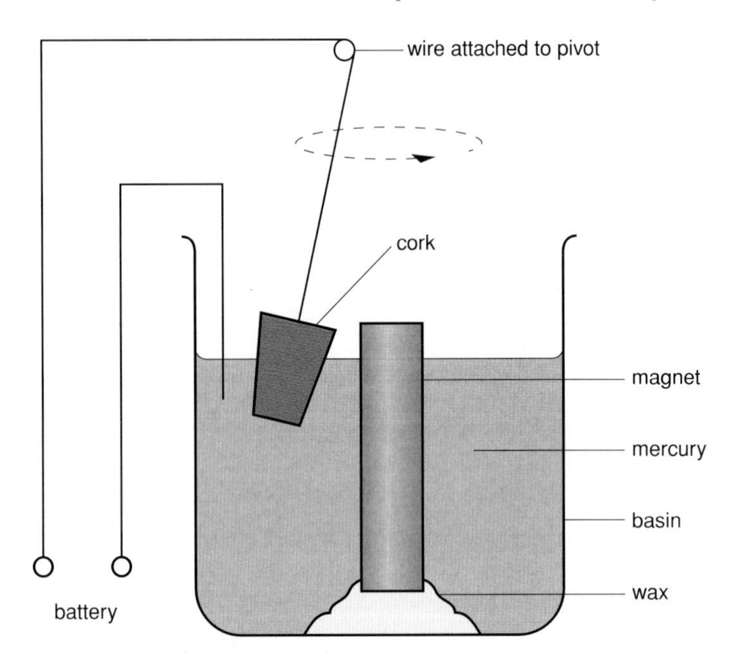

Did you know?
You can make a model motor like Faraday's using strong sodium chloride solution instead of mercury.

A moving wire

A wire will move in a magnetic field when a current flows through it. The picture shows how to investigate this.

A 'swing' has been made of copper wire. It hangs in a magnetic field.

When a current flows through the wire, it is pushed sideways. You can make the force bigger in two ways:

- increase the current;
- use stronger magnets.

If the current flows the other way, or if the magnets are reversed, the swing will be pushed in the opposite direction.

b How will the force change if the current is smaller?

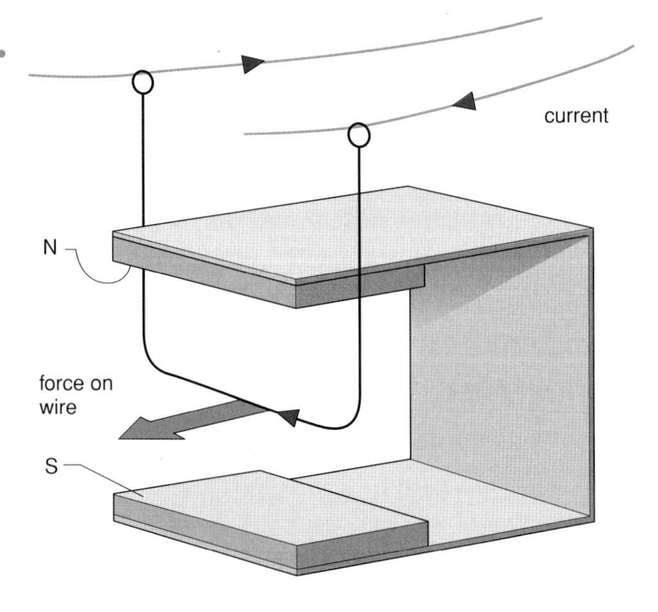

The model motor

Now we can look at the model motor again, and see a different way of explaining it.

- On one side of the coil, the current is flowing from front to back. The force on this side is upwards.
- On the other side of the coil, the current is flowing the other way, from back to front. The force on this side is downwards.

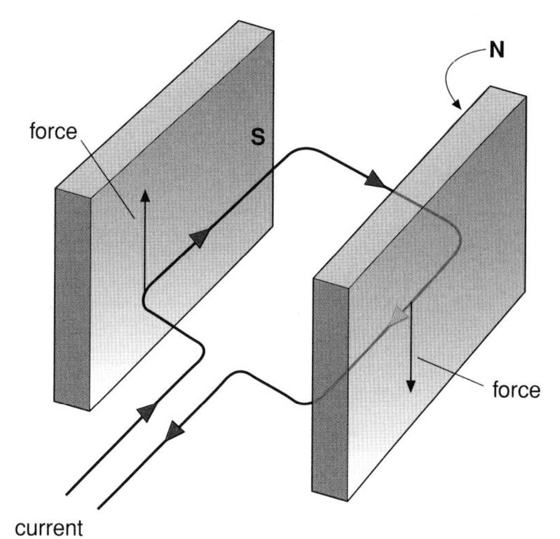

This pair of forces makes the motor spin around.

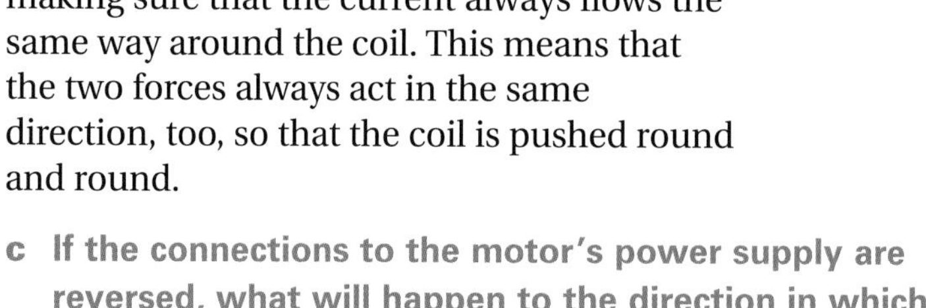

The brush connections are a clever way of making sure that the current always flows the same way around the coil. This means that the two forces always act in the same direction, too, so that the coil is pushed round and round.

c **If the connections to the motor's power supply are reversed, what will happen to the direction in which the current flows? Explain why the motor will turn in the opposite direction.**

Questions

1 **Look at the picture of the wire swing. The force on the wire can be changed in several ways. Copy the table, and tick the correct boxes to show how the force will change.**

Change made	Force increases?	Force reverses?
making a bigger current flow		
reversing the magnets		
using stronger magnets		
making the current flow in the opposite direction		

2 **Which of the things in this list are needed to make an electric motor?**

A coil of wire

An electric current

An ammeter

A magnetic field

A switch.

Summary

- When a wire carrying a current is placed in a magnetic field, there may be a force on it.
- The force can be made bigger by increasing the current and by increasing the strength of the magnetic field.
- Reversing the direction of the current or of the field reverses the direction of the force.

10:16 Making electricity

If you have a bicycle, it may have a **dynamo** fitted. As you travel along, the dynamo turns and generates electricity for the lights. A dynamo saves you the cost of batteries.

Each of these large red generators in a power station produces enough electricity to supply a small town.

A moving magnet

Inside a dynamo, a magnet spins around. As it spins past the coil, it produces a voltage between the ends of the coil. This voltage makes a current flow through the light.

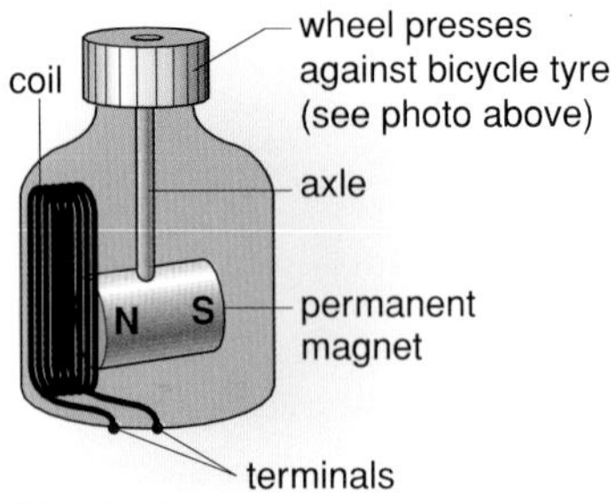

Bicycle dynamo.

A spinning coil

Some **generators** work like an electric motor in reverse. They have a coil of wire which is made to spin around in a magnetic field. A voltage is produced between the ends of the coil.

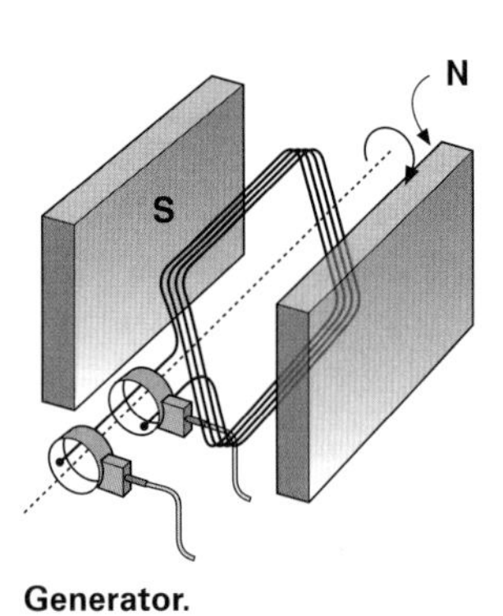

Generator.

a Copy the diagrams and write the words generator and motor in the correct boxes.

electricity → [ ] → movement

movement → [ ] → electricity

Investigating electricity generation

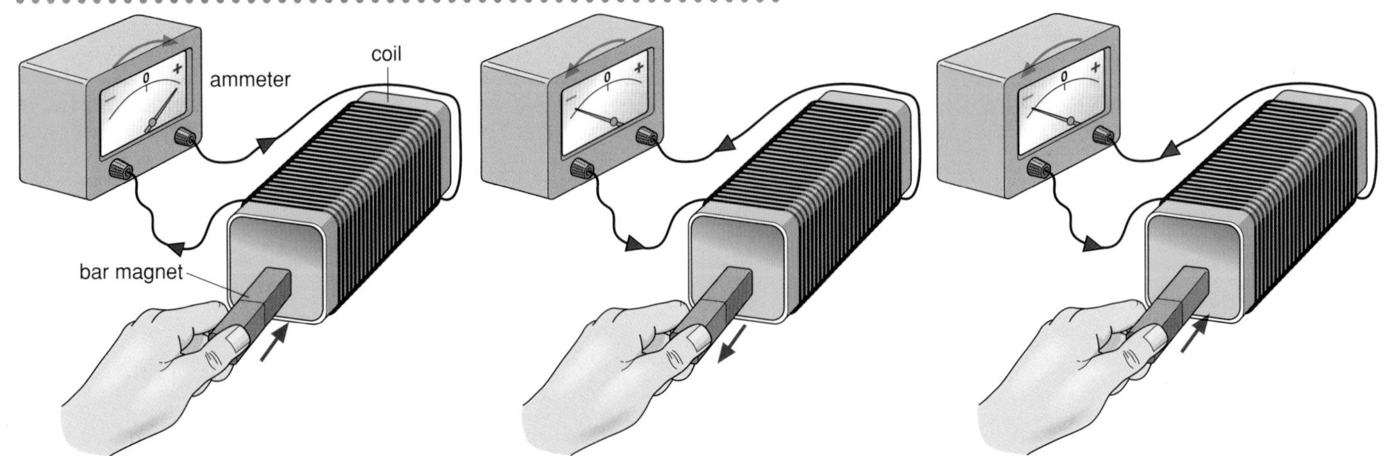

Push the magnet into the coil. The ammeter shows that a small current flows around the circuit.

Now pull the magnet out. The current flows in the opposite direction.

Try again with the other end of the magnet. The current flows the other way round.

If you hold the magnet still and move the coil, you will see the same effects. We say that:

- the coil *cuts through* the magnetic field;
- a voltage is *induced* between its ends.

b Which of these will change the direction of an induced current?

moving the magnet faster

moving the magnet in the opposite direction

using the opposite pole of the magnet

A bigger voltage

Experiments like these suggest four ways of increasing the p.d. induced in a spinning coil:

- spin the coil faster;
- use a stronger magnetic field;
- use a coil with more turns of wire;
- use a coil with a greater area.

c List four ways of *decreasing* the p.d. induced in a spinning coil.

Did you know?
A bicycle dynamo produces alternating current.

Questions

1 Copy and complete the following sentences, using words from the list to fill the gaps.

current voltage magnet coil circuit

In a dynamo, a __________ spins around inside a __________.

A __________ is produced between the ends of the coil.

A __________ will flow if the coil is part of a complete __________.

2 You can induce a voltage across a coil by moving a magnet into it, as shown. Which three of the following will produce a bigger voltage?

moving the magnet in the opposite direction

using a stronger magnet

using a coil with more turns

using the other end of the magnet

moving the magnet more quickly

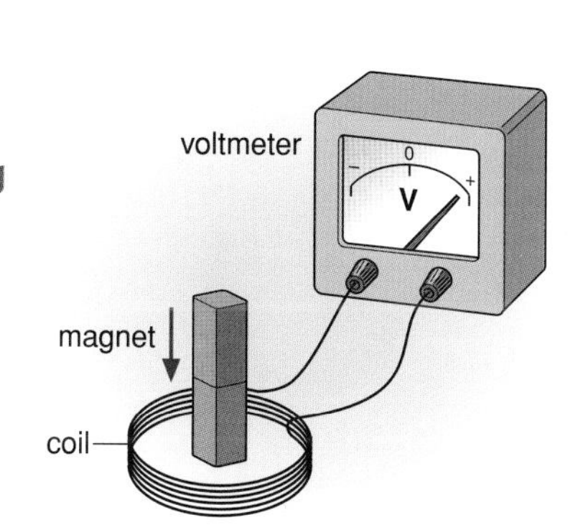

Summary

- In a generator, a coil of wire is made to turn in a magnetic field. This produces a p.d.
- If a wire is moved so that it cuts through a magnetic field, a p.d. is induced between its ends.
- The induced p.d. can be made bigger by:
  - moving the wire or coil more quickly;
  - using a stronger magnetic field;
  - using a coil with more turns;
  - using a coil with a bigger area.

10:17 Transformers

Giant pylons hold up the power lines which carry electricity around the country. The electricity is at a very high voltage, sometimes 400 000 V. This is very dangerous, so the pylons keep the wires safely above our heads.

These power lines carry electricity from power stations to where it is needed. They make up the National Grid.

Changing voltage

Transformers are used to change the voltage of the electricity supply. You may know where there are transformers near your house or school. Signs warn of the danger of the high-voltage electricity.

a Where have you seen electricity sub-stations like this? What warning signs did they have?

This electricity sub-station has transformers which reduce the voltage to the level needed by the consumers. People have been killed when they have entered sub-stations like this.

Around the country

Electricity is generated in power stations. Then it must be distributed around the country. Different consumers require different voltages.

The voltage is changed by transformers. They increase the voltage from the power station. Then they reduce it again before it reaches the consumers.

b Look at the diagram below. At what voltage is electricity generated in the power station?

At what voltage is it transmitted along the power lines?

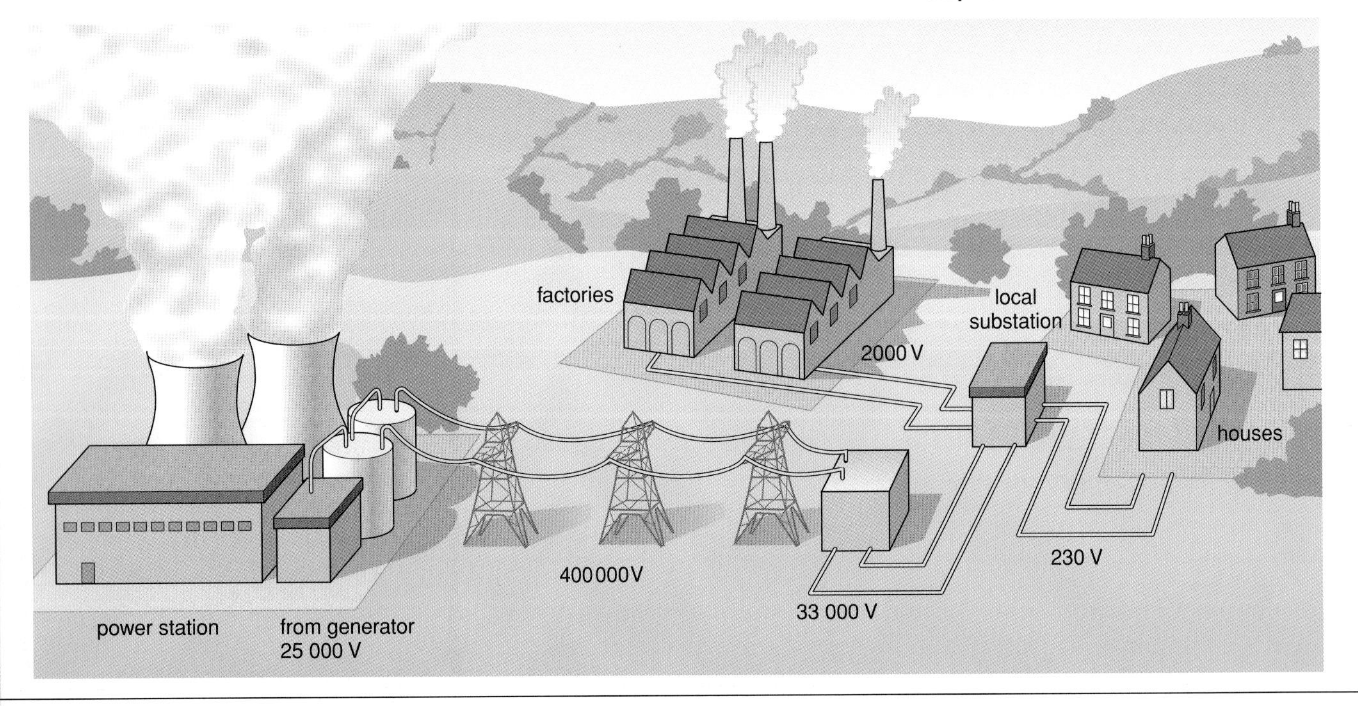

Transformers at home

The electricity supply to our homes is at 230 V. This voltage is too high for many of the things we use. They have transformers in them to reduce the voltage.

Mains electricity is an alternating (a.c.) supply. This means that the current flows back and forth in the wires; it doesn't flow steadily in one direction. Transformers only work with a.c. electricity.

c Why does the personal stereo in the drawing need a transformer if it is to be plugged into the mains electricity supply?

> **Did you know?**
> In France, some power lines carry electricity at over 1 million volts.

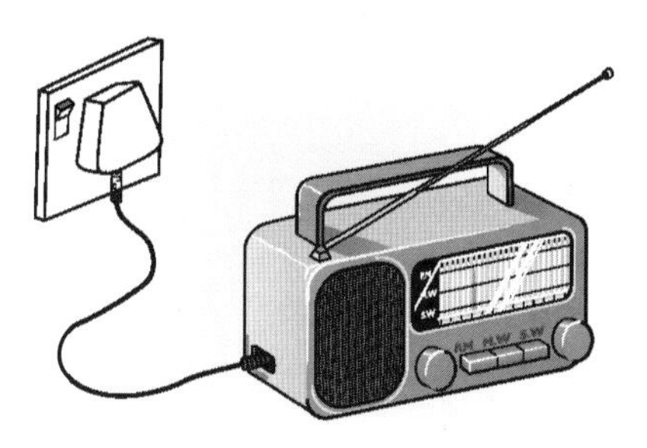

Radio plugged into mains.

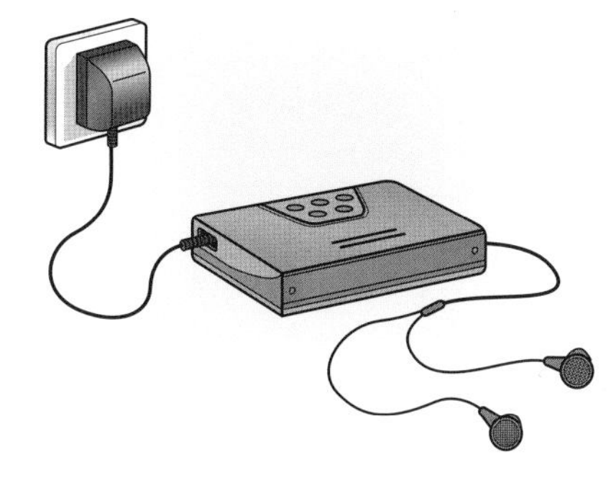

Walkman plugged into mains via a transformer.

Questions

1 Why are hazard warning signs displayed on electricity sub-stations?

2 Copy the boxes below. Choose words from the list to fill the boxes. You will have to use one of the words twice.

transformer **generator** **consumer** **power lines**

[ ] → [ ] → [ ] → [ ] → [ ]

3 Copy the sentences below, choosing the correct word from each pair.

At a power station, transformers increase/decrease the voltage of the supply before it is transmitted along the power lines.

At a sub-station, transformers increase/decrease the voltage of the supply before it reaches the consumer.

Summary

- Transformers change the voltage of an a.c. supply.
- At a power station, transformers are used to produce very high voltages.
- Then local transformers reduce the voltage to a safe level for use by consumers.

End of module questions

1 The diagram shows a three-pin plug for the electrical mains. Match words from the list with each of the labels 1–4 on the diagram.

earth pin
fuse
live pin
neutral pin

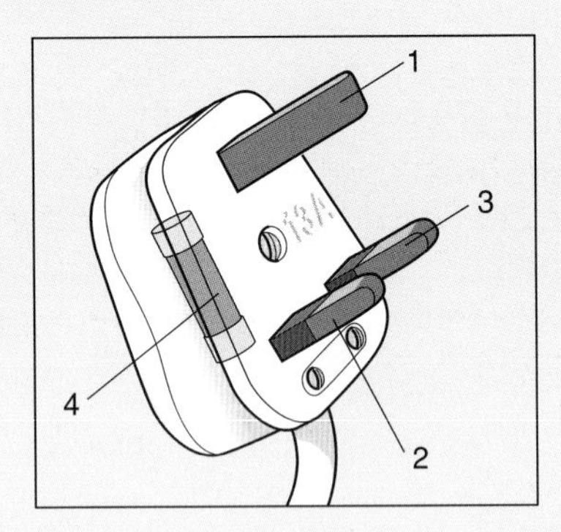

2 The sentences below describe how electrical measurements are made.

Choose words from the list for each of the spaces 1–4 in the sentences.

ammeter
voltmeter
in series
in parallel

The potential difference across a resistor is measured using a ___1___ .

The meter must be connected ___2___ with the resistor.

The current through the resistor is measured using a ___3___ .

The meter must be connected ___4___ with the resistor.

3 Components in a circuit are represented by circuit symbols.

Match words from the list with each of the components labelled 1–4 in the circuit diagram.

cell
filament lamp
diode
switch

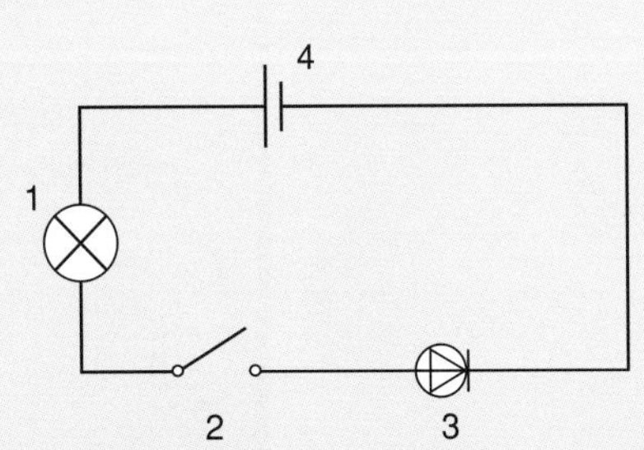

4 In an experiment, the following circuit was set up.

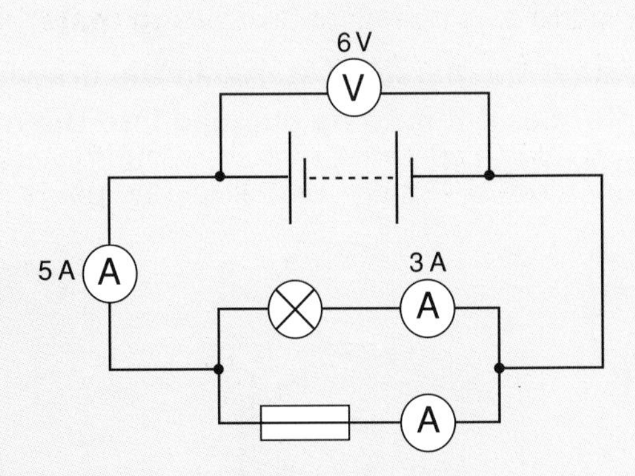

Match the readings from the list with each of the numbers 1–4 in the table.

2 A
3 A
5 A
6 V

1	the total voltage across the circuit
2	the total current flowing around the circuit
3	the current flowing through the lamp
4	the current flowing through the fuse

5 Match words from the list with each of the numbers 1–4 in the table.

electrodes
conductors
electrons
ions

1	substances through which electric current can flow
2	charged particles which carry electric current through a liquid
3	charged particles which carry electric current through a metal
4	terminals used to pass electric current through a liquid

6 Which **two** of the following devices make use of electromagnetic effects?

battery
filament lamp
loudspeaker
relay
switch

7 A pupil rubbed a plastic rod on a piece of cotton cloth. He found that a scrap of paper was attracted by the rod.

Which **two** of the following reasons might explain this?

the rod has been magnetised
the rod is electrically charged
the rod is made of an insulating material
the cloth is made of an insulating material
the paper is electrically charged

8 The diagram shows an electromagnet made by some students. It will pick up a few paper clips.

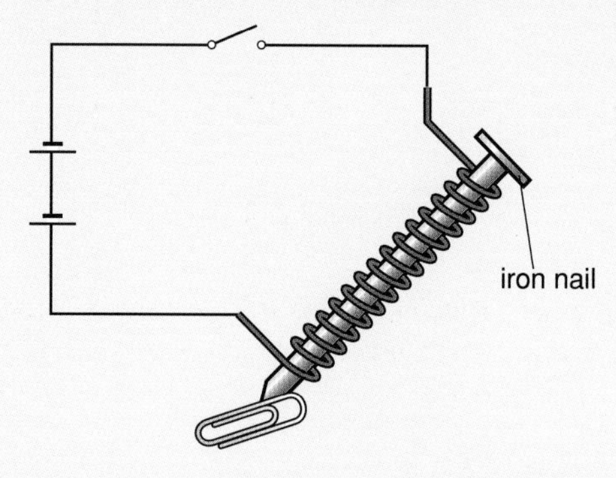

8.1 Why did the students use an iron nail?

A iron conducts electricity
B iron gets charged up
C iron is a magnetic material
D iron is a stiff material

8.2 The students wanted to make the electromagnet pick up more clips. What would make it stronger?

A use fewer turns of wire
B increase the current
C use insulated wire
D reverse the battery

8.3 What would make the magnet weaker?

A add a resistor in the circuit
B use an ammeter
C use more nails
D increase the voltage of the battery

8.4 When the students switched off the electromagnet, some paperclips remained sticking to the nail. Why was this?

A a small current still flowed in the coil
B the nail had become magnetised
C the nail had become electrically charged
D the clips had become electrically charged

9 Some pupils were investigating an electrical hairdryer.

9.1 The label showed that the hairdryer needed a current of 4 A. What value of fuse would be suitable?

A 3 A
B 4 A
C 5 A
D 13 A

9.2 The hairdryer became too hot. It switched itself off until it had cooled down. Then it started to work again. How did this happen?

A the fuse in the mains plug blew
B a circuit breaker in the hairdryer stopped the current flowing
C the current flowed away through the earth wire
D the mains voltage dropped to reduce the current

9.3 The fuse in the mains plug protects the circuit because

A it switches on and off easily
B it is connected to the earth wire
C it is easily replaced when it blows
D the fuse wire melts when too much current flows

9.4 On the hairdryer label, the students saw the term '50 Hz'. This means that

A the hairdryer can carry a current of 50 amperes
B mains electricity has a frequency of 50 cycles per second
C the safe voltage is 50 volts
D the power rating of the hairdryer is 50 watts

10 The diagram shows a wire connected to a meter. When a student moves the wire towards the magnet, the pointer on the meter moves to the right.

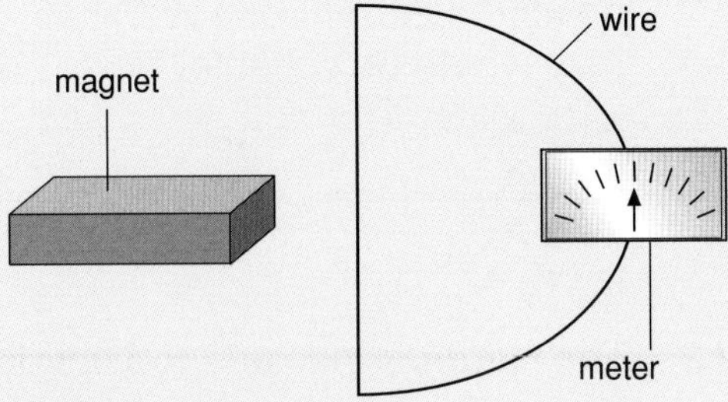

10.1 Why does a current flow through the wire?

A because the wire has become magnetised
B because the wire is attracted by the magnet
C because the wire is cutting through the magnetic field
D because the wire is repelled by the magnet

10.2 How could the student make the pointer on the meter move to the left?

A use a stronger magnet
B stop the wire moving
C move the wire backwards
D move the wire further forwards

10.3 How could the student make a bigger current flow in the wire?

A use a weaker magnet
B use the other end of the magnet
C move the wire more quickly
D use a more sensitive ammeter

10.4 Which of the following describes how a generator works?

A a magnet is rotated in a magnetic field
B a coil of wire is rotated in a magnetic field
C a magnet is moved into a coil of wire
D a coil of wire is rotated by an electric motor

Glossary of terms

acid A chemical that will react with a base to form a salt and water: weak acids have a sour taste, strong acids are corrosive.

acid rain Rain in which polluting substances such as acidic sulphur dioxide are dissolved.

addiction Taking drugs regularly with the result that it is difficult give them up.

aerobic respiration Chemical reactions in cells which use glucose and oxygen to release energy. Carbon dioxide and water are produced as waste products.

air passages Tubes which connect your mouth and nose to your lungs. Air moves through these tubes when you breathe in and out. The names of the air passages are trachea, bronchi and bronchioles.

alkali A base that dissolves in water: the solution has a pH greater than 7.

alkali metal A reactive metal from Group 1 of the periodic table, such as sodium, that reacts with water to form an alkaline solution.

alloy A mixture of two or more metals; e.g. brass is an alloy of copper and zinc.

alternating current (a.c.) Current which flows in one direction and then the other, constantly reversing.

alveoli The small air sacs in your lungs where oxygen diffuses into your bloodstream. Alveoli increase the surface area of your lungs which makes gas exchange more efficient.

anaerobic respiration The chemical reactions which release energy from glucose in the absence of oxygen. Lactic acid is produced as a waste product.

antibodies Chemicals produced by white blood cells which destroy bacteria and viruses.

antitoxins Chemicals produced by white blood cells which overcome the poisons (toxins) released by certain microbes.

aqueous solution A solution in water.

artery Blood vessel carrying blood away from the heart to other organs and tissues.

atmosphere The layer of gas that surrounds the Earth; approximately 80% nitrogen, 20% oxygen.

atom The smallest part of an element that still has the properties of that element.

atomic number See proton number.

atria The top two chambers of the heart. Blood flows from your veins into these chambers.

bacteria Very small single-celled organisms. Each bacterial cell has cytoplasm surrounded by a membrane, but has no distinct nucleus. Bacteria can reproduce very quickly and some cause disease.

base A metal oxide or metal hydroxide that reacts with an acid to give a salt plus water.

battery Two or more cells connected end-to-end to give a bigger p.d.

bauxite Aluminium oxide (Al_2O_3); the ore of aluminium.

beds (of rock) Layers of sedimentary rock – originally horizontal.

bile Fluid produced by the liver and stored in the gall bladder which assists in the digestion and absorption of fats.

biodegradable Something that can be broken down easily by natural, biological processes.

bladder An organ in the abdomen that stores urine.

blast furnace A special high-temperature furnace used to produce iron from iron ore by carbon reduction; the high temperature is produced by burning coke in a blast of hot air.

boiler The part of a power station where water is heated to provide high-pressure steam.

brain The part of the nervous system that co-ordinates most of the body's activities.

bronchioles The smallest tubes of the breathing system, ending in alveoli.

capillaries Blood vessels with a wall which is only one cell thick. Substances pass into and out of your blood in capillaries.

carbohydrases Enzymes which break down (digest) carbohydrates such as starch.

carbohydrates Foods, such as starch and sugars, which are the main energy-releasing foods in a diet.

carbon dioxide The gas produced as a waste product of respiration and absorbed during photosynthesis.

carbon monoxide A poisonous gas produced when carbon compounds burn in a limited supply of oxygen.

carbon reduction The removal of oxygen from a compound by displacement by carbon; e.g. iron oxide is reduced to iron by heating with carbon (coke).

catalyst Something that speeds up a chemical reaction without actually taking part in the reaction itself. Enzymes are examples of catalysts.

cell membrane The part of a cell that controls the entry and exit of materials.

cell sap The liquid that fills the vacuole in a plant cell.

cell wall The outer part of a plant cell, giving it shape and strength.

cellulose A carbohydrate, which strengthens the wall of a plant cell.

cement A powder which sets hard when mixed with water: made by heating limestone with clay.

chlorophyll A green pigment in plant cells that captures light energy for photosynthesis.

chloroplast Part of a plant cell that contains chlorophyll.

chromosomes Structures in a cell which look like long threads. The threads contain genes which control your inherited characteristics.

cilia Tiny strands which cover the lining of the nose and trachea. Their flicking movement carries dust and bacteria away from the lungs.

ciliary muscle A muscle in the eye that can alter the shape of the lens to focus objects.

circulatory system The system which pumps blood around the body. It is made up of blood vessels (arteries, veins and capillaries) and a pump (the heart).

clot When blood vessels are cut a clot is formed to seal the cut. This prevents loss of blood and stops bacteria getting into your body. The clot is the scab that forms over a cut or graze.

concrete An 'artificial rock' made by mixing gravel with wet cement: used as a building material.

condense Turn back from a gas to a liquid.

conduction The transfer of energy or electricity through a material, without the material itself moving.

continental drift The process where continents move over the surface of the Earth, carried on the moving plates.

convection currents Currents that form when a liquid or gas (or very hot rock) is heated from below. Energy is transferred as the material flows.

co-ordinator The part of the nervous system, either the brain or the spinal cord, where impulses pass from sensory neurones to motor neurones.

core: inner and outer The central part of the Earth made from iron and nickel; the outer part is liquid but the inner part is solid.

cornea The transparent part of the front of the eye, mainly responsible for focusing light on the retina.

corrosion A chemical reaction that attacks the surface of a metal, usually by water, acid or air.

cost-effective When you save more money than you spend, for example by fitting insulation.

coulomb (C) The unit of electric charge.

cracking The process by which long-chain hydrocarbons are broken up into shorter and more useful hydrocarbons.

crude oil A natural mixture of hydrocarbons found in some sedimentary rocks: formed from decayed animals and plants.

crust The thin, hard and brittle outer layer of the Earth.

cryolite A natural compound of aluminium used to lower the melting point of bauxite during electrolysis.

cuticle An outer skin on plant leaves, made of a waxy material that reduces evaporation of water.

cutting A part of a shoot, cut off and planted to produce a new plant.

cytoplasm The part of a cell where most chemical reactions occur.

denature To alter the structure of an enzyme so that it no longer works.

denitrifying bacteria Bacteria that get their energy by breaking down proteins and releasing nitrogen into the air.

dependent Unable to stop taking a drug.

diabetes A condition where the pancreas does not produce sufficient insulin, resulting in high blood sugar concentrations.

dialysis When a machine is used to treat kidney failure by taking over the functions of the diseased kidney.

diffuse The free electrons in a metal move around rapidly, at random. They spread out within the metal, carrying energy from hotter places to colder places.

diffusion Molecules of a gas are moving about all the time. Because of this movement the molecules spread themselves out evenly. This process is called diffusion. Diffusion involves movement of the molecules from a region of higher concentration to a region of lower concentration.

digestive system As food moves along your gut enzymes are released to break down large and insoluble food molecules into small soluble molecules. Your stomach, small intestine and pancreas are examples of organs which make up your digestive system.

dilate To get wider.

diode A device which allows current to flow in one direction only.

direct current (d.c.) Current which flows in one direction only.

displacement When a more reactive element pushes a less reactive element from its compound; e.g. iron will push copper out of copper sulphate solution.

distillation A method of purification; a liquid is boiled to form a gas, then cooled and condensed back to a liquid.

double circulatory system One side of your heart pumps blood around your lungs. The other side pumps blood around the rest of your body. This forms a double circulation.

dynamo A type of electrical generator.

earthquake A shock wave that passes through the Earth when rocks that have been under enormous pressure suddenly break and move along a fault line.

efficiency The fraction of the energy supplied to a device which is made good use of (transferred to useful forms).
electric charge A property of some particles which causes them to attract or repel each other. Can be positive or negative.
electrical energy Energy being transferred by electricity.
electrodes Rods used to conduct electric current in and out of a solution during electrolysis.
electrolysis Splitting a molten or dissolved chemical compound into simpler substances by passing an electric current through it.
electrons Negatively charged particles, smaller than an atom, which move through metals when a current flows.
element A substance made of one type of atom only.
emphysema A disease which destroys alveoli reducing their surface area. This means that people with emphysema cannot get enough oxygen into the blood.
emulsification The formation of thousands of tiny fat droplets from large fat globules. Bile produced by your liver emulsifies fats in this way. Tiny fat droplets can be more easily digested by lipase enzymes.
energy transfer When energy is passed from place to place, or from one object to another.
environment The surroundings of an organism.
enzymes Substances produced by cells which speed up reactions. For example, foods are broken down more rapidly by the action of digestive enzymes.
exchange surface A surface where materials such as gases enter or leave an organism.
extrusive rocks Igneous rocks formed when magma pours out onto the surface of the earth before it cools and sets, e.g. basalt lava.
faeces By the time food reaches your large intestine all the useful substances will have been absorbed leaving behind waste substances. In the large intestine water is absorbed from this waste to form faeces.
fats Food molecules which have a very high energy content. They are broken down in digestion to form fatty acids, glycerol and fat droplets.
fault line A crack in the crust of the Earth where the rock has broken and moved.
fibre Although fibre cannot be digested it forms an important part of a healthy diet by helping to form faeces. A lack of fibre can lead to diseases of the large intestine.
fossil fuel Coal, oil and gas; fuels made from the remains of decayed plants and animals found in some sedimentary rocks.
fossils Remains of once living things preserved in the rocks.
fractional distillation A form of distillation where only a partial separation of the liquids in a mixture is obtained.
fractionating tower A tall tower used in the industrial fractional distillation of crude oil; different liquids condense out at different levels.
fractions The different liquids produced from a complex mixture such as crude oil by fractional distillation.
gall bladder Vessel which stores bile after secretion from the liver, and before release into the intestine.
gene Information to control inherited characteristics is stored in the genes which are found on chromosomes in cells.
generator A device, often in a power station, which is turned and which produces electricity using electromagnetism.
geothermal power station A power station which makes use of the energy of hot rocks underground.
gland An organ that produces materials for use in other parts of the body.
global warming Increased proportions of greenhouse gases such as carbon dioxide are causing the average temperature at the surface of the Earth to rise.
glucagon A hormone, secreted by the pancreas, that increases the concentration of glucose in the blood.
gravitational potential energy (GPE) The energy which something has when it has been lifted upwards.
greenhouse effect The Earth is kept warm because gases (such as CO_2) in the atmosphere make it more difficult for heat energy to escape into space.
group Name given to the vertical columns of the periodic table: elements in the same group have similar properties.
heat energy Energy transferred from a hotter object to a colder one.
homeostasis Keeping conditions inside the body constant.
hormone A chemical, produced by one part of an organism, that controls a process in another part of the organism.
hydrocarbons Compounds made of carbon and hydrogen atoms only.
hydroelectric scheme Where water is stored behind a dam, so that it can be used to turn turbines and generate electricity.
igneous rock Rock formed from solidified magma (molten rock).

immunity By having white blood cells which produce antibodies you are able to destroy particular bacteria or viruses before they can cause a disease. Producing antibodies makes you immune to a particular disease.

infra red radiation Energy spreading out from a hot object in the form of waves.

insulator A material which is not a good conductor of heat energy or electricity.

insulin A hormone, secreted by the pancreas, that decreases the concentration of glucose in the blood.

intrusive rocks Igneous rocks formed when magma is forced into other rocks, and cools and sets within them, e.g. granite.

ion A charged particle made from one or more atoms.

ionic compound A compound formed by the electrostatic attraction between oppositely charged ions.

ions Electrically charged particles; an atom or group of atoms with an electrical charge.

iris The coloured part of the eye that controls the amount of light entering the eye by altering the width of the pupil.

joule (J) The scientific unit of energy.

kidney An organ that gets rid of waste materials and controls the water and salt content of the blood.

kinetic energy The energy which any moving object has.

lactic acid Lactic acid is produced when glucose is broken down in the absence of oxygen during anaerobic respiration.

large intestine The large intestine is the final part of your gut. This is where water is absorbed into the bloodstream and faeces are formed.

lava Magma that pours out onto the surface of the Earth.

leaf Part of a plant whose main function is to photosynthesise (manufacture food).

lens Part of the eye that helps to focus light on the retina.

light-dependent resistor (LDR) A device whose resistance decreases when light shines on it.

light energy Energy you can see, spreading out from a very hot object.

limiting factor A factor such as light intensity that limits the rate of a process such as photosynthesis.

lipase Lipases are fat-digesting enzymes. Fats and oils are also known as lipids.

lithosphere The thin crust attached to a thicker slab of upper mantle rock.

liver An organ responsible for many processes including the production of bile, the breakdown of amino acids into urea and the storage of sugars as glycogen.

magma Molten rock.

magnetic poles The Earth's magnetic poles are the points close to the geographic poles from which the Earth's magnetic field seems to come.

mantle The middle, soft, rocky layer of the Earth that can move very slowly.

metamorphic rock A rock that has been changed and recrystallised by heat and pressure (but has not actually melted).

microbe Microbes are very tiny organisms which can only be seen under a microscope. Bacteria and viruses are examples of microbes.

mucus Mucus is a slimy material produced by the body for lubrication. For example, mucus prevents food damaging the lining of the gut as it gets pushed along.

muscle An organ which brings about movement.

negative charge See electric charge.

nerve Collection of neurones held together in a bundle by connective tissue.

neurone A nerve cell.

neutral Neither acid nor alkali; 7 on the pH scale.

neutralisation The process by which an acid and an alkali (or other base) may 'cancel out' to give a salt plus water.

nitrates Ions needed by plants to make amino acids.

non-renewable Once a non-renewable energy source has been used up, it can never be used again.

nucleus (of atom) The central part of the atom containing the proton(s) and, for all except hydrogen, the neutrons.

nucleus (of cell) The part of a cell which contains genetic information.

ohm (Ω) The unit of electrical resistance.

optic nerve A nerve which carries information from the eye to the brain.

ore A natural compound of a metal from which the metal can be extracted.

organ Organs are different tissues working together to carry out a job in your body. For example, muscle and glandular tissues form the stomach which is an organ in the digestive system.

organ system A number of organs working together form an organ system. For example, the stomach, pancreas and small intestine are organs in the digestive system.

osmosis The net movement of water molecules through a partially permeable membrane from a region of high concentration of water molecules to a region of lower concentration.

oxygen debt Oxygen is needed to remove lactic acid produced in anaerobic respiration. The oxygen that is needed is called an oxygen debt.

palisade mesophyll Cells under the upper epidermis of a leaf, where most photosynthesis occurs.

pancreas An organ near the stomach which secretes both digestive enzymes and hormones concerned with the control of blood sugar.

partially permeable Allowing small molecules to pass through quickly, but larger molecules more slowly.

pay-back period The time taken to recover the initial cost of insulation from the savings made.

periodic table A way of arranging all the different elements in a table to link up those with similar properties.

phloem A plant tissue responsible largely for transporting food materials around the plant.

phosphorus An element required for good plant growth.

photosynthesis A process taking place in green parts of a plant. Water and carbon dioxide react together in sunlight in the presence of chlorophyll to produce sugars and oxygen gas.

pH scale A 0–14 scale used to measure from strong acid (0) through neutral (7) to strong alkali (14).

plasma Plasma is the liquid part of blood. Many substances such as sugars, urea and carbon dioxide are transported in plasma.

plastics Polymers; very long-chained molecules made by joining lots of small molecules together.

poly(ethene) A long-chained polymer made by joining lots of ethene molecules together in an addition reaction.

polymers Very long-chained molecules made by joining lots of small molecules together.

positive charge See electric charge.

potential difference (p.d.) The voltage between two points which can make an electric current flow between them.

power The rate at which energy is transferred.

protease Proteases are enzymes which digest proteins to produce amino acids.

proteins Proteins are a very important group of food substances. Much of your body is made from protein. You make your body protein from the amino acids absorbed from your gut.

proton A sub-atomic particle with a positive charge and a relative mass of 1.

proton number The number of protons in an atom.

pulse A pulse is formed as the heart pumps blood along arteries. When the heart beats it forces blood along arteries causing them to swell. This can be felt as a pulse in various parts of your body such as in your wrist.

pumped-storage scheme Where water is pumped uphill, ready to be released through a turbine to generate electricity when there is a sudden high demand.

pupil The hole surrounded by the iris in the eye.

quicklime Calcium oxide (CaO) made by heating limestone.

radiation Energy spreading out from a source (e.g. infra red or light energy), or carried by particles (e.g. from a radioactive substance).

raw material A natural material such as rock which can be used to make other products.

reactivity series Common elements (mostly metal) arranged in order of reactivity; this order stays the same for all reactions.

receptor A cell which is sensitive to a stimulus.

reduction The removal of oxygen from a compound, e.g. iron oxide is reduced to iron: reduction may also be defined as the gain of electrons.

reflex action An automatic response to a stimulus.

relative atomic mass The number of protons added to the number of neutrons in an atom.

reliable source (of electricity) A steady source of electricity which is available over long periods of time.

resistance The greater the resistance of a component, the smaller the current that will flow through it for a given p.d.

respire To release energy from food materials.

response Behaviour of an organism when stimulated.

retina The inner layer of the eye, containing light-sensitive cells.

ripple marks Fossilized ripples preserved in sediment that has been shaped by moving water.

rock cycle The natural process whereby rocks can change from igneous through sedimentary to metamorphic and back to igneous.

root The organ of a plant that anchors it in the ground and takes in water and mineral ions.

root hairs Cells in roots largely responsible for the uptake of water and mineral ions.

rotary kiln A special kiln used to make cement. It rotates to mix the raw material and stop the cement from clumping as it is formed.

sacrificial protection A method used to protect steel from rusting in highly corrosive environments (such as the sea) by bolting on a more reactive metal such as zinc or magnesium; the more reactive metal corrodes instead of the steel.

salt One of the products of neutralisation; an ionic compound such as sodium chloride or copper sulphate.

sclera The outer, white part of the eye.

sedimentary rock Rock usually made from fragments of other rocks or shells which are laid down as beds of sediment, often in the sea.

sense organ Receptor that is sensitive to specific stimuli.

sensitivity The ability to react to changes in the environment.

sieve tubes The structures in phloem that transport sugars.

skin Your skin acts as a protective covering preventing entry of harmful bacteria and viruses.

slaked lime Calcium hydroxide ($Ca(OH)_2$): made by adding water to quicklime.

small intestine Portion of the alimentary canal between the stomach and large intestine which plays a major role in the digestion and absorption of food.

solvents Liquids used to dissolve substances – if sniffed they can cause damage to the body and become addictive.

sound energy Energy which you can hear, travelling as sound waves.

stem The organ in a plant that supports the leaves and transports materials.

stimulus (plural: **stimuli**) A change in the environment which can be detected by organisms.

stoma (plural: **stomata**) A tiny hole in the surface of a leaf that allows gases to enter and exit.

surface area All the outside of an organism.

suspensory ligament A structure which holds the lens of the eye in place.

sweat A liquid produced by glands in the skin. It cools the body as it evaporates.

tectonic plates The rigid slabs of lithosphere that move about the surface of the Earth, driven by the slow convection currents in the mantle below.

thermal decomposition Breaking down a chemical compound by heating.

thermal energy Another name for heat energy.

thermistor A device whose resistance decreases as its temperature increases.

thermit A solid displacement reaction in which a more reactive metal pushes a less reactive metal from its oxide, e.g. aluminium pushes iron from iron oxide.

tissue Cells which carry out the same job in your body group together into a tissue. For example, muscle cells group together to form a muscle tissue.

toxin A toxin is a poisonous substance. When microbes infect the body they may release toxins which make you ill.

transformer A device used to increase or decrease the voltage (p.d.) of an electricity supply.

transition metals The 'everyday' metals such as iron and copper, found in the central block of the periodic table.

transpiration The loss of water vapour from the shoot of a plant.

turbines The part in a power station that is made to spin around, and which turns the generator.

turgor The pressure exerted by the contents of a plant cell on its cell wall.

universal indicator A mixture of dyes which changes colour with different pH values.

unreliable source (of electricity) A source of electricity which varies a lot, so that it is not always available.

urea Urea is a waste product produced in the liver and removed from the body by the kidneys.

urine A waste fluid, produced by the kidneys, containing urea, excess water and excess salts.

vaccinated People are vaccinated by introducing mild or dead microbes into their bodies. Their white blood cells then produce antibodies making them immune to a particular disease.

vacuole A fluid-filled sac in most plant cells.

variegated A leaf which has green areas containing chlorophyll and yellow/white areas with no chlorophyll.

veins Blood vessels which carry blood back to the heart from body tissues.

ventilation The movement of air in and out of the lungs is an example of ventilation. Breathing air in gets oxygen into the alveoli. Breathing out removes waste carbon dioxide from the body.

ventricles The bottom two chambers of the heart are the ventricles. They have thick muscular walls to pump blood to the lungs and to the rest of the body.

villi The lining of the small intestine is covered with thousands of tiny folds called villi. The villi increase the surface area of the lining which makes the absorption of food more efficient.

viruses Microbes that are smaller than bacteria. They have a protein coat surrounding a few genes. They can only reproduce inside living cells.

viscous Thick, not runny; high-viscosity liquids (viscous liquids) are thick, low-viscosity liquids are runny.

volatile A volatile liquid will turn into a gas (form a vapour) very easily.
voltage A measure of how much energy is transferred by a cell or supply into each coulomb of charge. It is measured in volts (V).
watt (W) The unit of power. 1 watt is 1 joule per second.
wax A shiny material that is impermeable to water.
weeds Unwanted plants growing in cultivated plots.
white blood cells White blood cells form part of the body's defence system against microbes which cause disease. They do this by ingesting microbes, and by producing antibodies and antitoxins. Unlike red blood cells, white blood cells contain a nucleus.
wilting The drooping of a plant through lack of water.
wind farm A collection of wind generators.
wind generator A machine which is made to spin round by the wind and which generates electricity.
withdrawal symptoms What a person feels when they stop using an addictive drug.
xylem A plant tissue largely responsible for the transport of water in the plant.
xylem vessels The structures in xylem that transport most of the water in plants.

Note: Some of the terms used in this book have more than one recognised spelling. For internet searches it is worth using both spellings separately, in order to get the best results. Words that this applies to are foetus (fetus) and sulphur (sulfur).

Data sheets

Reactivity series of metals

Potassium	most reactive ↑
Sodium	
Calcium	
Magnesium	
Aluminium	
Carbon	
Zinc	
Iron	
Tin	
Lead	
Hydrogen	
Copper	
Silver	
Gold	
Platinum	↓ least reactive

(elements in italics, though non-metals, have been included for comparison).

Formulae of some common ions

Positive ions

Name	Formula
Hydrogen	H^+
Sodium	Na^+
Silver	Ag^+
Potassium	K^+
Lithium	Li^+
Ammonium	NH_4^+
Barium	Ba^{2+}
Calcium	Ca^{2+}
Copper(II)	Cu^{2+}
Magnesium	Mg^{2+}
Zinc	Zn^{2+}
Lead	Pb^{2+}
Iron(II)	Fe^{2+}
Iron(III)	Fe^{3+}
Aluminium	Al^{3+}

Negative ions

Name	Formula
Chloride	Cl^-
Bromide	Br^-
Fluoride	F^-
Iodide	I^-
Hydroxide	OH^-
Nitrate	NO_3^-
Oxide	O^{2-}
Sulphide	S^{2-}
Sulphate	SO_4^{2-}
Carbonate	CO_3^{2-}

The periodic table of elements

KEY

Mass number A	1
	H Hydrogen
Atomic number (Proton number) Z	1

1	2											3	4	5	6	7	0
																	4 **He** Helium 2
7 **Li** Lithium 3	9 **Be** Beryllium 4											11 **B** Boron 5	12 **C** Carbon 6	14 **N** Nitrogen 7	16 **O** Oxygen 8	19 **F** Fluorine 9	20 **Ne** Neon 10
23 **Na** Sodium 11	24 **Mg** Magnesium 12											27 **Al** Aluminium 13	28 **Si** Silicon 14	31 **P** Phosphorous 15	32 **S** Sulphur 16	35 **Cl** Chlorine 17	40 **Ar** Argon 18
39 **K** Potassium 19	40 **Ca** Calcium 20	45 **Sc** Scandium 21	48 **Ti** Titanium 22	51 **V** Vanadium 23	52 **Cr** Chromium 24	55 **Mn** Manganese 25	56 **Fe** Iron 26	59 **Co** Cobalt 27	59 **Ni** Nickel 28	63 **Cu** Copper 29	64 **Zn** Zinc 30	70 **Ga** Gallium 31	73 **Ge** Germanium 32	75 **As** Arsenic 33	79 **Se** Selenium 34	80 **Br** Bromine 35	84 **Kr** Krypton 36
85 **Rb** Rubidium 37	88 **Sr** Strontium 38	89 **Y** Yttrium 39	91 **Zr** Zirconium 40	93 **Nb** Niobium 41	96 **Mo** Molybdenum 42	99 **Tc** Technetium 43	101 **Ru** Ruthenium 44	103 **Rh** Rhodium 45	106 **Pd** Palladium 46	108 **Ag** Silver 47	112 **Cd** Cadmium 48	115 **In** Indium 49	119 **Sn** Tin 50	122 **Sb** Antimony 51	128 **Te** Tellurium 52	127 **I** Iodine 53	131 **Xe** Xenon 54
133 **Cs** Caesium 55	137 **Ba** Barium 56	139 **La** Lanthanum 57	178 **Hf** Hafnium 72	181 **Ta** Tantalum 73	184 **W** Tungsten 74	186 **Re** Rhenium 75	190 **Os** Osmium 76	192 **Ir** Iridium 77	195 **Pt** Platinum 78	197 **Au** Gold 79	202 **Hg** Mercury 80	204 **Tl** Thallium 81	207 **Pb** Lead 82	209 **Bi** Bismuth 83	210 **Po** Polonium 84	210 **At** Astatine 85	222 **Rn** Radon 86
223 **Fr** Francium 87	226 **Ra** Radium 88	227 **Ac** Actinium 89															

Elements 58–71 and 90–103 have been omitted.

The value used for mass number is normally that of the commonest isotope, e.g. ^{35}Cl not ^{37}Cl.

Bromine is approximately equal proportions of ^{79}Br and ^{81}Br.

Formulae list

This list shows the formulae for quantitative relationships in the Physical Processes section of the specification which candidates will be expected to recall (N.B. for convenience, formulae are also given here in symbolic form even though this form is not required by the specification).

potential difference (volt, V) = current (ampere, A) × resistance (ohm, Ω) $V = IR$

power (watt, W) = potential difference (volt, V) × current (ampere, A) $P = VI$

energy transferred (kilowatt hour, kWh) = power (kilowatt, W) × time (hour, h) $E = Pt$

total cost = number of Units × cost per Unit

energy transferred (joule, J) = power (watt, W) × time (second, s) $E = Pt$

acceleration (metre/second squared, m/s²) = $\frac{\text{change in velocity (metre/second, m/s)}}{\text{time taken for change (second, s)}}$ $a = \frac{v - u}{t}$

wave speed (metre/second, m/s) = frequency (hertz, Hz) × wavelength (metre, m) $v = f\lambda$

efficiency = $\frac{\text{useful energy transferred by device}}{\text{total energy supplied to device}}$

work done = energy transferred

work done (joule, J) = force applied (newton, N) × distance moved in direction of force (metre, m) $W = Fs$

power (watt, W) = $\frac{\text{work done (joule, J)}}{\text{time taken (second, s)}}$ $p = \frac{W}{t}$

weight (newton, N) = mass (kilogram, kg) × gravitational field strength (newton/kilogram, N/kg) $w = mg$

change in gravitational potential *energy (joule, J)* = weight *(newton, N)* × change in vertical *height (metre, m)* $gpe = mg\Delta h$

kinetic energy (joule, J) = $\frac{1}{2}$ × mass (kilogram, kg) × speed² [(metre/second)², (m/s)²] $ke = \frac{1}{2} mv^2$

Index

Note: Glossary entries are not included in the index.